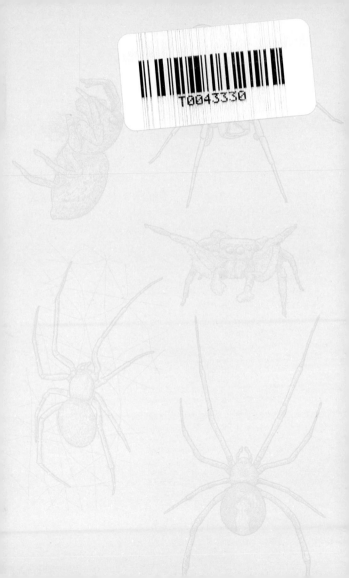

THE LITTLE BOOK OF
SPIDERS

THE LITTLE BOOK OF
SPIDERS

With color illustrations by Tugce Okay

SIMON D. POLLARD

PRINCETON UNIVERSITY PRESS
PRINCETON AND OXFORD

Published in 2024 by Princeton University Press
41 William Street, Princeton, New Jersey 08540
99 Banbury Road, Oxford OX2 6JX
press.princeton.edu

Copyright © 2024 by UniPress Books Limited
www.unipressbooks.com

Library of Congress Control Number 2023943775
ISBN 978-0-691-25182-0
Ebook ISBN 978-0-691-25181-3

Typeset in Calluna and Futura PT

Printed and bound in China
1 3 5 7 9 10 8 6 4 2

British Library Cataloging-in-Publication Data is available

This book was conceived, designed, and produced by UniPress Books Limited

Publisher: Nigel Browning
Managing editor: Slav Todorov
Project development and management: Ruth Patrick
Design and art direction: Lindsey Johns
Copy editor: Caroline West
Color illustrations: Tugce Okay
Line illustrations: Ian Durneen

IMAGE CREDITS:
Alamy Stock Photo: 23 Wirestock, Inc.; 116 Tonia Graves; 131 Anton Sorokin.
Dreamstime.com: 17 Kacpura; 74 Sleepyhobbit; 87 Brett Hondow.
Nature Picture Library: 12, 90, 148 Emanuele Biggi; 29 Kim Taylor;
55 Michael Hutchinson. **Shutterstock**: 41 Sebastian Janicki; 95 khlungcenter;
119 Tobias Hauke; 126 LFRabanedo; 139 Couanon Julien. **Other**: 35 NASA;
47, 58, 62, 70, 105 Simon D. Pollard; 106 FranciscoJavierCoradoR;
146 Fiona Cross. **Additional illustration references**: 11 Marianne Collins;
45 Robert R. Jackson; 53 pxfuel.com; 73 Ken-ichi Ueda; 89 Alexandre S.
Michelotto; 99, 103 Simon D. Pollard; 111 benjamynweil; 112 Keizo Takasuka;
15 jtweed; 133 George Hachey; 151 deeqld.

Also available in this series:

THE LITTLE BOOK OF
BEETLES

THE LITTLE BOOK OF
BUTTERFLIES

THE LITTLE BOOK OF
TREES

Coming soon:

THE LITTLE BOOK OF
FUNGI

THE LITTLE BOOK OF
WEATHER

THE LITTLE BOOK OF
DINOSAURS

THE LITTLE BOOK OF
WHALES

CONTENTS

INTRODUCTION

When I was five years old and growing up in Christchurch on New Zealand's South Island, I had an unusual question for my uncle, Jim Pollard. Jim was an animal psychologist at the University of Canterbury who answered my many questions about the natural world living in my parent's garden. "Do bees think in English?", I asked. I thought all animals thought in English (with a kiwi accent!). He told me they didn't, and a couple of years and many questions later, I decided I wanted to study animals like my uncle. I never wanted to do anything else, and a decade later I started studying zoology at the University of Canterbury. In my third year in an entomology course, it was expanded to include spiders and a light turned on in my head. I wanted to become a spider biologist.

THE ARRIVAL OF SPIDER-MAN

The following year, Robert Jackson from the United States, whose field was animal behavior, joined the Zoology Department. Robert specialized in spider behavior and I became his apprentice as an undergraduate and postgraduate student. He has had a profound influence on my life and I am so grateful to have such a valued colleague and friend.

After completing my PhD I held post-doctoral fellowships at the University of Virginia and University of Alberta in Canada. While I enjoyed research, I also wanted to write about my research for a wide audience and so began a long association with *Natural History Magazine*, then based in the American Museum of Natural History in New York.

My first story was on my PhD and post-doc research on crab spiders and came out in October 1993. It was called "Little Murders." My latest story came out in the February 2022 issue and was about my research on an Asian crab spider that lives in pitcher plants. It was called "Living in a Death Trap." I feel as if I have been paraphrasing Agatha Christie book titles for my natural history story titles!

AMAZING SPIDER STORIES

The Little Book of Spiders brings together my career as a spider biologist, writer, and science communicator. I have loved researching the various spider stories included in the book and thank the numerous scientists who have helped to unravel the lives of spiders with innovative techniques and technologies.

Who could have imagined a spider the size of a grain of rice that can snap its jaws 800 times faster than you can blink (a species of *Zearchaea*); or spiders (species of *Portia*) that hunt other spiders, have the cunning of a mammal, can count, and have eyesight one-sixth as good as ours. Some small orb web spiders even build a giant replica of a spider in their web (species of *Cyclosa*). The decoy has eight legs and is thought to fool predators into thinking that it is a spider too big to attack.

We share our planet with around 50,000 different species of spider that are found living almost everywhere. I hope this book gives readers a new appreciation of these wonderful creatures, which live among us and mean us no harm. The tales they spin often seem beyond our imagination and the joy those stories bring adds to our appreciation of life on Earth.

Simon D. Pollard

ARACHNID ANCESTORS

The evolutionary journey that led to arthropods, a phylum that includes spiders, insects, and crustaceans—arguably the most successful group of animals on Earth—began with a human-sized ancestor that looked a bit like a lobster. It swam in the Earth's oceans from around 520 million years ago and had a body plan that paved the way for arthropods to move from sea to land. While this ancestor was known from fossils, they were flattened in formation and many unanswered questions remained. The relatively recent discovery of a remarkably well-preserved and unflattened 480-million-year-old fossil in Morocco allowed scientists to fill in the gaps.

FLEXIBLE BODY PARTS

Scientists now understand how this ancestor of spiders led to descendants that were able to evolve into so many different forms and occupy almost every possible part of the planet. The building blocks for this remarkable evolutionary success were a segmented body and segmented legs.

↓ A remarkably intact 480-million-year-old fossil of an arthropod ancestor answered many questions about the group's evolution.

↓ Two specimens discovered in amber in Myanmar show a 100-million-year-old spider ancestor with a whip-like tail.

→ Living in oceans 480 million years ago was a human-sized early ancestor of today's crustaceans, insects, and spiders. Named *Aegirocassis benmoulae* and looking like a lobster, it filtered seawater for plankton, like some whales today.

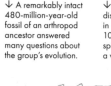

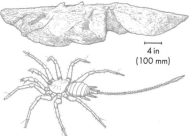

4 in
(100 mm)

II

ROCK FOSSIL SPIDERS

For every spider fossil found in rocks, about a thousand insect fossils are discovered. This is because spiders are soft-bodied and do not have the hard exoskeleton of insects, which makes them much less likely to leave fossilized remains.

SPIDER JIGSAW PUZZLES

Dr. John Nudds (of the University of Manchester) and Professor Paul Selden (of the University of Kansas) said that understanding the evolution of spiders from fossils is like having a jigsaw puzzle with most of the pieces missing and no picture of the puzzle on the box. With modern scanning techniques, such as computed tomography (CT) scanning, scientists have been able to analyze the relatively few spider fossils found to understand the early evolution of this incredibly successful group.

REFLECTING EYES

Two rock fossils from around 110 million years ago, found in shale in South Korea, are from an extinct family of night-hunting spiders, the Lagonomegopidae. These spiders lived from around 359 to 299 million years ago and were previously only known from amber fossils. Remarkably, fossilization preserved a structure found in both fossils, at the back of two of the spiders' large eyes. Called a tapetum lucidum, this reflects light that has passed through the retina back onto the retina, giving better vision at night. Just like the shine seen when you aim a flashlight into the eyes of a cat at night, the eyes of these fossils shone back at the researchers who discovered them.

BURROWING FOR 300 MILLION YEARS

The Mesothelae are a suborder of spiders and the sister group of all spiders. They lived around 300 million years ago and had a segmented abdomen. Their spinnerets (silk-spinning organs) were located in the middle of the abdomen (rather than at the back, as is the case for all spiders today). Remarkably, a single family (the Liphistiidae) in this group is still alive today and appears to closely resemble their distant ancestors. They live in burrows in forests and caves in Southeast Asia and Japan.

~ The first true spiders ~

While the mesothelids produced silk, it was only the evolution of spinnerets at the end of the abdomen that gave spiders much finer control over how they could use silk. This paved the way for their success. Fossils of the first true spiders have been dated to around 250 million years ago.

← Waiting for the vibrations of walking prey, a female Malaysian trapdoor spider, *Liphistius malayanus* (Liphistiidae), sits at the entrance of her burrow.

LIFE ON LAND

A round 100 species of burrow-dwelling spiders (Liphistiidae) from the ancient suborder Mesothelae survive today. The remaining 50,000 known species of spiders belong to two groups, the mygalomorphs and the araneomorphs, with the latter represented by around 47,000 different species. Of the 3,000 species of mygalomorphs, around 1,000 species are tarantulas. They tend to be big and hairy with the stocky body shape characteristic of mygalomorphs. One of the reasons for the success of the araneomorphs, compared to the mygalomorphs, is the orientation of their chelicerae (jaws).

FLEXIBLE FANGS

The chelicerae of mygalomorphs point downward, with the fangs folded away like the blade of a pocketknife. When the spider wants to use its fangs to strike at prey or defend itself, it has to raise its body slightly in order to open its fangs from their resting position. It also needs to be on a solid surface like the ground or the bark of a tree before it can stab downward with its fangs.

GOLIATH BIRD-EATING TARANTULA

Theraphosa blondi (Theraphosidae) is the largest spider on Earth and lives in burrows in the ground of rainforests in northern South America. It can weigh around 6 oz (180 g) and have a leg span of 12 in (30 cm) and a body length of 5 in (13 cm). It eats mostly large, ground-dwelling insects, worms, and amphibians, and very rarely birds. Its reputation for eating birds came from an illustration in a book published on natural history in 1705. The very talented Maria Sibylla Merian, after a trip to South America, had made an engraving of the spider eating a hummingbird. The Goliath bird-eating tarantula is a common name that has stuck for over 300 years.

Araneomorph spiders have fangs that move from side to side, so the spider can use them like a pair of pincers. This means they can catch prey while suspended in webs off the ground.

~ Not just for stabbing ~

Almost all spiders use their fangs to inject venom into prey to paralyze or kill it. The fangs of some spiders are used for tasks totally unrelated to predation. For example, trapdoor spiders use them to excavate soil and build burrows. Nursery web and spitting spiders carry their egg sacs by holding them with their fangs. Some male ant-mimicking jumping spiders have enlarged chelicerae and fangs, which are used in contests of strength with other males.

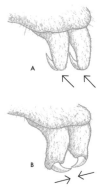

↑ The two main groups of spiders have fangs that work in different ways. (A) Mygalomorphs strike up and down while (B) Araneomorphs strike from side to side.

FOSSILS IN AMBER

Spiders entombed and preserved in liquid resin, which over time hardens into amber, offer a glimpse into their behavior from around one hundred million years ago. Fossils from rock almost always appear ancient, but fossils in resin can look as fresh as if they were inside a just-frozen ice cube.

SILK PRODUCTION

Spiders are known to have produced silk from around 400 million years ago, but the oldest spider silk was found inside a piece of 140-million-year-old amber on a beach in Sussex in the United Kingdom. While it came from an extinct ancestor of the familiar orb web spiders living today, its silk also contained the sticky droplets used to trap flying insects. Scientists suggest this may provide an insight into the evolutionary arms race as flying insects took to the skies and spiders evolved aerial webs to trap them.

A MOTHER IN AMBER

Today, female spiders from many different families show maternal behavior, which includes guarding their egg sacs and newly hatched spiderlings. Although it was accepted that this behavior would have a long evolutionary history, until recently there was no physical evidence. The discovery of a couple of fossilized female spiders with eggs and spiderlings in 100-million-year-old amber in Myanmar shows how long this behavior has been around. The females were guarding their offspring when they were smothered by flowing tree resin, which hardened over time and became amber. Researchers using CT scanning could identify the almost hatched spiderlings within the egg sac as well as the hatched spiderlings staying close to their mother.

↑ Trapped in Baltic amber, a 40-million-year-old spider provides a fascinating snapshot of how spiders evolved during this time.

WEB WITH PREY

The oldest known web holding trapped prey was found in a piece of amber from Spain and dated at 110 million years old. Tangled among the 26 strands of silk were the remains of a fly, beetle, wasp, and tiny mite.

SPIDER WITH PREY

Inside a piece of 100-million-year-old amber from Myanmar is a remarkable piece of preserved spider history. Engulfed by flowing resin was a juvenile social orb weaving spider as it attacked a parasitic wasp. Also trapped in the amber was a male, providing the first fossil evidence of social spiders.

AERIAL WEBS

T he evolution of flying insects led to the evolution of spider webs that could trap this out-of-reach prey. It seems two distantly related groups of spiders (the superfamilies Deinopoidea and Araneoidea) evolved different types of orb webs to trap flying insects. The Deinopoidea had horizontal orb webs with thousands of fine silk fibers wrapped around larger fibers. The Araneoidea had the familiar vertical orb web with sticky droplets attached to dry silk.

LOSING ORB WEBS

The sticky orb web is often considered the pinnacle of spider web evolution. Recent research has shown that many families of spiders no longer use orb webs and evolved different ways of capturing prey. While orb webs are effective at capturing insects in flight, they also make the spider waiting in the web vulnerable to predators like birds. Dr. Charles Griswold and colleagues, from the California Academy of Sciences, consider that "the orb web has been an evolutionary base camp rather than a summit."

↓ Sticky glue droplets on the spiral silken threads of orb webs hold prey until it is captured by the spider.

↓ Many orb web spiders have decorative silken structures called stabilimenta, which may stop birds from flying into the web.

→ A female giant golden silk orb-weaver, *Nephila pilipes* (Nephilidae), sits in her enormous web and waits for prey. Recent research has revealed that the yellow markings on the spider's legs attract flying insects during the day and moths at night.

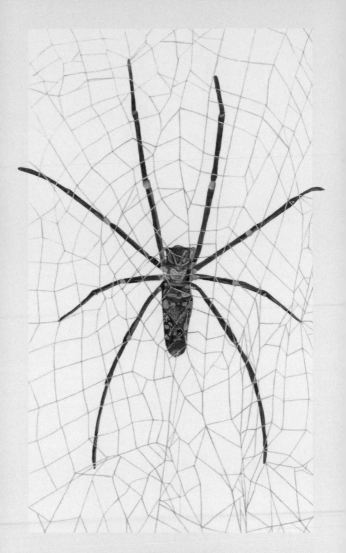

SPIDERS ARE EVERYWHERE

Spiders are found living in practically every habitat on Earth, from high in the Himalayas to beneath the waves along coastlines. They can survive freezing temperatures in the Arctic and high temperatures in deserts. By overwintering inside silken nests under a blanket of snow they can survive temperatures of -40°F (-40°C). The hottest place where spiders have been found is Death Valley, in California, in an area where a ground temperature of almost 140°F (60°C) has been recorded. A range of species were found, including web builders, crab spiders, and jumping spiders. In tropical and subtropical rainforests spiders can be found in huge numbers represented by a diverse range of species and lifestyles.

LIVING IN CAVES WITHOUT EYES

Around a thousand different spider species have been found living in caves worldwide. Some of these species have lost all of their eight eyes. This may have been selected for due to the resources needed to maintain a sense organ that was no longer of any use to the spider.

UNDER THE SEA

In the intertidal zone along the coastlines of New Zealand and New Caledonia lives the marine spider *Desis marina* (Desidae). During low tide it hunts on dry land for small crustaceans, before retreating to its silken nest. When the spider senses a rising tide, it seals its retreat, so it can survive being submerged beneath the waves. Remarkably, depending on the tidal range, the retreat is large enough to provide the spider with an air supply for almost three weeks. If it is submerged for this long, it can lower its respiratory rate to ensure it does not run out of air. It can also survive lower oxygen levels compared to other spiders if its air supply is running low.

SPIDERS WITH ALTITUDE

In 1924, a naturalist on an expedition to the Himalayas discovered jumping spiders living at 22,000 ft (6,000 m). Since there was no vegetation or apparent prey for the spiders he concluded that they must survive by cannibalism. Obviously, if this were the case, the population would disappear very quickly indeed. In 1961, another naturalist also discovered the same jumping spiders living at high altitudes and found that they were, in fact, living on flies blown up by winds from lower altitudes. The Himalayan jumping spider, *Euophrys omnisuperstes* (Salticidae), is also found living at much lower altitudes, but it holds the record for living at the highest known altitude for any spider. Its species name means "standing above everything."

Also, many spiders have small eyes and these may have been easier to lose over time. None of the highly visual jumping spiders have been found living in caves. Another feature of some spiders living in caves is the evolution of very long legs to sense a world in darkness, like the long antennae of some cave-dwelling insects.

SPIDERS IN ANTARCTICA

Antarctica is the only continent where spiders do not live, but a few spiders have visited. They traveled as stowaways with cargo being transported to the frozen continent. It ended up being a one-way trip for the spiders since they were all found dead from the extreme cold.

→ Often spiders that live in caves have lost their eyes and instead rely on other senses in order to perceive their dark world.

SPIDER BRAINS

Spiders have two main parts to their brains, which lie close to the eyes. The top part, the supraoesophageal ganglion, is connected to the lower part, the suboesophageal ganglion, and both parts control responses to internal and external sensory information. Nerves from both ganglia send and receive information from the various organ systems within the abdomen.

VERY BRAINY SPIDERS

The complex and intelligent behavior of species of *Portia*, a jumping spider that eats other spiders, is changing the way researchers think about what a brain that could sit comfortably on the head of a pin is capable of. Jumping spiders have excellent eyesight—that is, about one-sixth as good as ours—and researchers found that they were able to watch movies on tiny screens (see Chapter 9, page 104). This has led researchers to ponder whether its extraordinary eyesight influenced how its brain evolved and gave it the intelligence of a small mammal.

TINY BRAIN POWER

The number of nerve cells that make up the brain of a tiny web-building spider—*Mysmena* species (Mysmenidae)—are not fewer or smaller than those in the brain of the large, web-building spider *Trichonephila clavipes*, a golden silk orb-weaver, even though this spider weighs about 20,000 times more. Building an orb web requires the same brain power, whether the spider is big or small. While *Trichonephila* can easily fits its brain into its body, *Mysmena*'s brain fills about 80 percent of its body and even around 25 percent of the space in its legs.

→ *Trichonephila clavipes* feeds on a large insect that became trapped in its web.

RECORDINGS FROM A SPIDER'S BRAIN

Over many years researchers have learned how jumping spiders, with their exceptional eyesight, see the world. Six lower resolution eyes act as motion-detectors and direct what the two high-resolution eyes should focus on, whether it's prey, mates, or predators. Until recently, understanding how a jumping spider's brain, which is about the size of a sesame seed, processes this information has been a mystery. Since their bodies are under pressure, trying to insert an electrode into the brain was like sticking a needle into a fluid-filled balloon. However, researchers were able to insert the super-strong, super-thin electrodes used on mammals into the spider's brain and begin to understand how the brain processes information in these very clever animals.

A SKELETON ON THE OUTSIDE

Spiders have a skeleton on the outside (called an exoskeleton), which has two body segments. These are joined by a narrow circular waist called a pedicel. The front segment, the cephalothorax (fused head and thorax), has a hard, armor-like covering and is where the legs are attached to the body. It is also used as an attachment surface for muscles, just like the bones in our bodies. The front of the cephalothorax has the chelicerae (jaws) with their fangs. Either side of the chelicerae are the short, leg-like pedipalps, which are used to manipulate prey. In mature males the pedipalps carry sperm.

The abdomen is less rigid and more flexible, so it can expand after the spider has eaten. The pedicel connects the gut, nervous, and circulatory systems from the cephalothorax to and from the abdomen. At the end of the abdomen are silk-spinning organs, which are finger-like projections known as spinnerets.

JOINTED LEGS

All spiders have eight jointed legs, with each leg having seven segments. At the end of each leg, spiders have two or three tarsal claws. Web-building spiders have three claws, with the middle claw being used to manipulate silken threads when the spider is building a web. Spiders with two claws often have adhesive hairs (called scopulate hairs). These branch into hundreds of smaller hairs and allow the spider to climb smooth, vertical walls.

← The molted exoskeleton of a jumping spider. This image also shows the molted lenses of four of the spider's eight eyes.

LOSING A LEG

Spiders have a mechanism by which they can deliberately amputate one or more of their legs. Called autotomy, it allows the spider to possibly escape a predator that has grabbed one of its legs. Also, if a spider is stung in the leg by a stinging insect, like a wasp or bee, it can quickly autotomize the leg before the venom reaches its body. The first two leg joints attached to the spider's body are the coxa and trochanter, which have a joint membrane between them. Under tension the membrane shears and the leg detaches. Muscles in the coxa pull the remains of the membrane inward and coagulated blood seals the wound to stop excessive blood loss.

~ Sensory hairs ~

A spider's body, especially the legs, are covered in hundreds and even tens of thousands of different sensory hairs that can detect ground and airborne vibrations and chemical cues. The last segment of a spider's leg is the tarsus and this contains the tarsal organ, which can sense temperature and humidity.

GROWING BIGGER

Since spiders have a rigid exoskeleton, they need to replace this with a larger version as they grow bigger. This is known as molting and it is a risky undertaking. The spider grows a new exoskeleton beneath the old one. Increasing blood pressure within the cephalothorax causes the top of the old exoskeleton to detach. The spider then extracts itself from the old exoskeleton and again increases blood pressure until the new one reaches full size. The new exoskeleton hardens over a few hours. If the spider has lost one or more legs before molting it can, over several molts, grow new legs. If it loses legs after its final molt, it cannot replace them.

HOW SPIDERS BREATHE

J ust like humans, spiders require a constant supply of oxygen to breathe. On the underside of the abdomen are two small openings where air enters the book lungs. These resemble a very unusual book (hence the name) with inflated, blood-filled pages and air pockets in between. The blood picks up oxygen from the air pockets, which is then transported to the heart to be pumped around the spider's body. At the same time oxygen is taken up, carbon dioxide diffuses out of the book lungs via the abdominal opening.

BREATHING THROUGH A TUBE

As well as a pair of book lungs, most spiders, except for ancient groups, including tarantulas, also breathe through a highly branched and efficient system of tubes known as tracheae. Depending on the spider, these may have one or two openings called spiracles and are located near the rear of the abdomen. Some small spiders only breathe using their tracheae.

↓ The crab spider *Misumena nepenthicola* (Thomisidae) lives inside pitcher plants and uses a bubble of air to breathe while it is submerged in the plant's fluid, where it remains to hunt and hide.

→ The diving bell spider, *Argyroneta aquatica* (Dictynidae), is the only spider known to spend almost its entire life underwater. These spiders come to the surface to attach air bubbles to their abdomen and legs, which keeps their diving bell full of air.

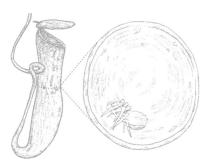

HEART AND CIRCULATION

Spider blood (hemolymph) carries blood and nutrients around the body. It's also the fluid that drives the hydraulic mechanism essential for walking. Contracting muscles in the cephalothorax to reduce the size of its body, the spider increases blood pressure and forces fluid into the legs, which straighten, and then leg muscles controlled by nerves and the central nervous system initiate walking.

THE HEART

Hemolymph is pumped around the body by a muscular tubular heart in the abdomen. The heart has a large artery called the anterior aorta, which takes blood into the cephalothorax, and a posterior aorta that supplies blood to the abdomen. When the heart contracts, blood is forced around the spider's body via the major arteries, which branch into smaller arteries. Valves at both ends of the heart close after contraction to stop blood flowing back into the heart.

CIRCULATION

While spider blood vessels branch into smaller ones, hemolymph eventually leaves the vessels and blood bathes the spider's tissues, supplying them with oxygen and nutrients. This is called an open circulatory system. When the heart relaxes, hemolymph flows back into the abdomen and the book lungs where carbon dioxide diffuses out. The hemolymph picks up more oxygen before returning to the heart to be pumped around the body again.

~ Blue blood ~

The oxygen-carrying molecule in our bodies is called hemoglobin, which contains iron and makes our blood red. In spiders, the molecule carrying oxygen to the tissues is hemocyanin, which contains copper and gives spider blood a faint blue color.

JUMPING JUMPING SPIDERS

Spiders walk by changing the internal pressure of their body fluids to increase and decrease the amount of fluid in the legs. This is coordinated with the contraction and relaxation of leg muscles. Some spiders can jump short distances, but some jumping spiders can jump almost forty times their own body length. The maximum distance for a human long jump is around five body lengths. To get ready to jump, the spider raises its front pair of legs and attaches a safety line of silk. Then it suddenly increases the amount of fluid pushed into either the third or fourth pair of legs, or both, and becomes airborne. Due to the spider's excellent eyesight it can land on target.

↓ While attached to a safety line of silk, jumping spider *Marpissa muscosa* (Salticidae) leaps from one acorn to another.

FLUID FEEDERS

Spiders have a liquid diet and particles larger than one-thousandth of 0.00004 inches (one micron) are filtered out and do not enter the spider's digestive system. Once prey have been paralyzed or killed (usually with venom), or wrapped in silk, spiders begin feeding. Some spiders crush prey with their jaws, which often have tooth-like projections to help break open the insect's exoskeleton and expose the tissues. Other spiders feed through tiny holes in the prey made by their fangs. When these spiders have finished feeding, the prey is left intact, except for their sucked-out insides.

EATING NOXIOUS PREY

The woodlouse-eating spider, *Dysdera crocata* (Dysderidae), has an impressively large set of fangs. These allow the spider to hold its main prey, woodlice (or pill bugs as they are known in the United States), in a scissor-like grip with one fang resting on top. The other fang injects venom into the prey's soft underside. Very few spiders eat woodlice because they secrete a noxious chemical along the edge of their body. By holding the woodlouse away from its own body the spider can avoid the prey's chemical defenses. When the spider feeds, it carefully uses a fang to tear a hole in the center of the woodlouse's underside, away from the defensive droplets, and mixes digestive fluid with the prey's tissues before sucking up the liquid.

→ The red-colored *Dysdera crocata* is originally from the Mediterranean, but it now has a cosmopolitan distribution.

A SILKEN BAG OF SOUP

Uloborid spiders, known as hackled orb web spiders, do not have venom glands and instead wrap their prey in silk, so it cannot escape. As they wrap, they compress the prey, so it can be squeezed into a small bag of silk. The spider then regurgitates digestive fluid into the bag. This dissolves soft membranes like leg joints and allows the spider to extract the prey's tissues, which leak out after being compressed. The spider's mouthparts do not make contact with the prey during feeding and it sucks the partially digested nutrients through the silk wrapping, which also acts as a filter to stop any solid parts of the prey interfering with feeding.

SOUP STRAINERS

Spiders start digesting prey outside their own digestive system, by effectively vomiting onto or inside the prey's body. They secrete a digestive fluid, which contains various enzymes, and this is mixed with the prey's tissues to break them down. Once the prey is liquid enough, the spider sucks the liquid into their internal digestive system. An initial filter of hairs at the front of the tiny mouth acts like a soup strainer. Just inside the mouth in the pharynx is a narrow, tooth-like, lined slit that also strains out larger particles.

~ Pumping stomach ~

To get the partially digested nutrients into the midgut where digestion is completed, spiders use a pumping stomach. This draws fluid in and pumps it into the midgut through the contraction and relaxation of various groups of muscles. At each end of the stomach are valves that ensure fluid moves in the direction required at the time. Once digestion is complete, nutrients are moved around the body and waste is filtered and water conserved by the malpighian tubules, which have a similar function to our kidneys.

HUGE APPETITE

Spiders can eat an enormous amount of food in a short space of time. This explains why they can survive for many months without feeding. The spider's midgut, which extends from the stomach in the cephalothorax and into the abdomen, is where digestion takes place. After a huge meal, the midgut expands and squeezes into most of the spaces inside the spider's body. In the abdomen it expands around the silk glands and the female's reproductive organs.

BRAIN SQUEEZE

In the cephalothorax the midgut splits into two extensions on either side of the stomach. Each of these extensions has numerous finger-like projections. As the midgut fills with food, these projections extend down into the first joint (coxa) of the spider's eight legs. They also extend forward and expand around the brain. In jumping spiders, the two long eye tubes are also pushed around by the food-filled projections.

↓ The jumping spider *Hyllus semicupreus* (Salticidae) from India uses silk for support and venom to catch prey like large grasshoppers.

↓ The orb web spider *Argiope submaronica* (Araneidae) from Costa Rica catches and eats proboscis bats (*Rhynchonyeteris naso*).

→ A female raft spider, *Dolomedes fimbriatus* (Pisauridae), eating a three-spined stickleback (*Gasterosteus aculeatus*). The spider catches the fish by resting its front legs on the water surface and then sensing the vibrations made by its prey as it swims. It then goes underwater and grabs the prey.

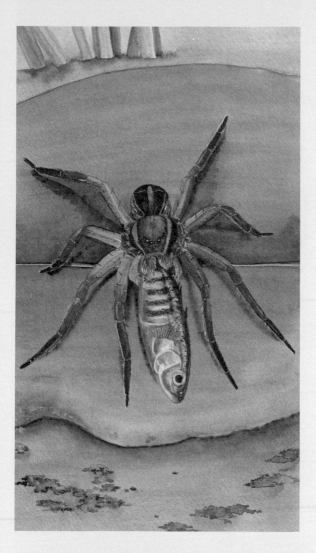

KEEPING UPRIGHT

In 1973, four years after people first walked on the moon, two common North American orb web spiders *Araneus diadematus* (Araneidae) were taken to the first NASA space station, Skylab, to see how zero gravity affected web building.

ORB WEBS AND GRAVITY

Most orb web spiders build asymmetric webs, with the hub closer to the top of the web. This means the prey-capture area below the hub is bigger than the one above it. The spider sits in the hub, facing downward, since gravity makes it easier and quicker to run down the web to capture prey than to run up the web. It is thought lyriform sense organs located in the waist between the cephalothorax and the abdomen can sense gravity, so the spider knows down from up. When the spider is facing downward, the pull of gravity causes slits in the lyriform organs to narrow and these signal the spider's central nervous system as to which way it is facing.

MUTINY ON THE SPACE STATION

In 2008, two more orb web spiders were taken to the International Space Station. One spider was able to spin a web in its sealed cage. The other spider was a backup in case the first spider died. Unfortunately, the backup spider managed to get into the other spider's cage. The two spiders competed for space as they both tried and failed to build proper orb webs. The fruit flies bred as spider prey also escaped to where the spiders were now living. After a month, the viewing window was obscured with a curtain of fly larvae smearing a nutrient medium and fly pupae. Obviously, the experiment didn't go as planned.

SPIDERS IN SPACE

The results from the first space spiders in 1973 and a trip to the International Space Station in 2008 gave mixed results because of difficulties in housing and recording the spiders' behavior. A successful trip to the space station in 2011, however, showed that the spiders built more symmetric webs in zero gravity than they did on Earth. The spiders were photographed every five minutes for eleven months, giving researchers a very detailed picture of how they behaved. They found that when the spiders had a light source above them, they used this, instead of gravity, to orient and build similar asymmetric webs as they did on Earth.

↑ In 2012, after a hundred days aboard the International Space Station, a Johnson jumping spider, *Phidippus johnsoni* (Salticidae), returned safely to Earth.

"LISTENING" IN WEBS

Prey and predators have always been locked in an ongoing evolutionary battle of wits, where prey don't want to be eaten and predators want to eat them. A brilliant example is spiders as predators and moths as prey. Moths have evolved a clever trick to avoid becoming prey when they fly into an orb web. They can lose sacrificial wing and body scales and slide between the sticky radial threads of the web, and so escape unharmed.

SPIDER VERSUS MOTH

Some orb web spiders have evolved an ingenious way of modifying their orb webs so they can trap moths. *Scoloderus cordatus* (Araneidae) from North America builds a vertical ladder web, with the familiar-looking orb web at the bottom. If moths fly into the silken ladder, they tumble downward and are stripped of their protective scales before being trapped at the bottom of the sticky orb web. The spider picks up vibrations through the silk from the tumbling victim, then rushes out and catches the moth as it becomes stuck in the orb web at the bottom of the ladder. Webs sprinkled with moth scales are testimony to their fate as the main prey of a moth-trapping web.

↓ Common European garden spider, *Araneus diadematus* (Araneidae), sits and waits in its web with its claws resting and "listening" out on the silken lines.

SPRING-LOADED TRAPS

The common house spider, *Parasteatoda tepidariorum* (Theridiidae), is often found in places where people live. It builds a messy-looking, tangled cobweb that traps flying insects. It also traps prey in a manner that is almost the opposite of how ladder spiders catch moths. It attaches sticky silken lines to the ground beneath its web. These lines are under tension and work as spring-loaded traps. If walking prey stick to them, the silk detaches and hoists them into the air. Since the prey are off the ground they hang helplessly. Silk-guided vibrations from their struggles alert the spider to the arrival of trapped prey and they are hoisted into the tangle web and bitten. Having silken lines under tension pulls on the tangle web above them and the web has extra-strong silk to attach the web and hold it in place.

~ Hoisting lizards ~

The spring-loaded traps are so effective they can even catch huge prey such as lizards and large insects that may weigh up to fifty times the weight of the spider. The spider uses a type of silken pulley system to hoist the prey off the ground and into the web, before having an enormous meal. One reason why lizards may stumble into the web and become prey is that they are attracted to the sound of trapped flies beating their wings in an attempt to escape from the web.

"LISTENING" ON THE GROUND

The weight of an insect walking on a leaf causes it to move around 0.000000004 in (0.00001 mm). Remarkably, bromeliad spiders— species of *Cupiennius* (Theridiidae) can pick up this movement from almost 6 ft (2 m) away. While the spider can also pick up airborne vibrations, it has specialized sense organs called lyriform organs on its legs that are sensitive to vibrations on the ground. Lyriform organs consist of numerous slits. Vibrations cause these slits to be slightly squeezed, which sends sensory information to the spider's brain for processing.

SENSING PREY WALKING

When a bromeliad spider is "listening," it raises its body and spreads its legs evenly, with the end of its leg (tarsus) resting on a surface like a leaf. Lyriform organs on the joint between the metatarsus and tarsus can pick up the vibrations from walking prey and tell the spider when it is close enough to grab.

↓ The lyriform organs (B) are sensitive to pressure and located on the spider's leg joint between the metatarsus (A) and the tarsus (C). They are able to pick up vibrations from prey walking nearby.

→ Sitting and waiting in a characteristic pose, this female tiger bromeliad spider, *Cupiennius salei* (Trechaleidae), picks up the vibrations of prey walking nearby. The spider hides during the day and hunts at night where it preys on insects, frogs, and lizards.

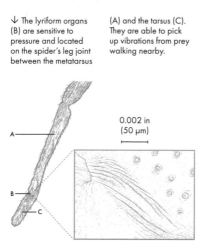

0.002 in (50 μm)

"LISTENING" IN THE AIR

For an animal of their size, jumping spiders have extraordinarily good eyesight and see as well as some primates. Recently, researchers used tiny micro-electrodes implanted in the brain of the North American jumping spider *Phidippus audax* (Salticidae). They wanted to identify the part of the brain used to process the visual information coming from its two big, forward-facing eyes, which allow these spiders to see as well as they do.

"LISTENING" TO BRAINS

When a researcher stood up, their chair squeaked on the floor and the spider's brain responded. As part of the recording device, when the electrodes in the spider's brain picked up a signal from the firing of nearby nerve cells, it caused a "pop" sound in a speaker that the researchers could hear. Soon, the researchers were clapping their hands from 10–15 ft (3–5 m) away and with each clap came a pop sound from the speaker. It was thought spiders can only "hear" sounds nearby and those far away are only audible to an animal with a tympanic membrane. They found the frequencies to which the spider responded matched the sound of the wing beats of parasitoid wasps, a jumping spider enemy.

~ Jumping spiders can hear ~

Previous research had shown that jumping spiders freeze when they hear parasitoid wasps flying nearby. Sensory hairs on the spider's legs are moved by sound vibrations and send signals to its brain. It is amazing to think jumping spiders may be able to hear you talking from across a room.

SPIDER HEARS A VIOLIN

The special sensory hairs on the legs of spiders that can pick up airborne vibrations are called trichobothria. This was discovered in the late 19th century by a spider specialist who was looking at the hairs on a spider's leg. He noticed that the hairs moved in response to the sound coming from a violin being played nearby. The researchers who discovered that jumping spiders could hear sounds from much farther away than previously thought have also been looking to see whether fishing and wolf spiders can hear sounds from a distance too. Since all spiders have trichobothria it is not surprising that they can. Talk to spiders in different tones and they may respond.

↓ Although jumping spiders are famous for their eyesight, the North American jumping spider *Phidippus audax* showed researchers that they are able to hear sounds very well too.

AMAZING EYES

Although most spiders have eight eyes, the majority have poor eyesight and rely on vibrations, touch, and chemical cues to perceive their world. The exceptions are exceptional, from jumping spiders with tiny telescopes inside their heads to net-casting spiders with the equivalent of night-vision goggles (see Chapter 7, page 92).

JUMPING SPIDER EYES

Jumping spiders are unusual because, unlike the majority of spiders, they are active during the day and rely on their excellent vision to find prey and mates and to avoid predators. From the outside the spider's most distinctive feature are the two large, forward-facing eyes with their convex lenses. Behind these lenses extend two long eye tubes.

~ Incredible vision ~

Just like our own camera-like eyes, the lenses in the principal eyes of jumping spiders focus images on a retina. In jumping spiders the retina lies at the end of the eye tube and is made up of four different layers. Due to the length of the eye tube and a second lens close to the retina, the eyes work a bit like a telescope and make distant objects appear closer. The first three layers of the retina allow the spider to see different colors, while the rear layer has the highest number of light-sensitive receptors and allows the spider to see details about one-sixth as good as our own eyes.

→ A jumping spider's two large, telescope-like eye tubes combined with the six smaller eyes means that it has excellent eyesight.

FLY TIGERS

Hundreds of years ago, the Chinese called jumping spiders "fly tigers" because they hunt prey like a tiger. Researchers have similarly called them "eight-legged cats," and when you watch a jumping spider stalk prey, it is easy to see why. While no other spiders, or any animals of a similar size, have eyesight to rival that of a jumping spider, wolf spiders and crab spiders both have reasonable eyesight and can see objects they are close to. Male wolf spiders even wave their front legs at females during courtship. Crab spiders can see nearby prey and will grab it with their long front legs.

~ Tiny retinas ~

Since the image captured by the front lenses is so much bigger than what can be captured by the relatively tiny retina, the spider has muscles around the eye tubes and can move them around to scan what the lenses are looking at. Each retina is shaped like a boomerang, the center of which has the highest density of photoreceptors. This is similar to the fovea centralis of our own eyes, which allows us to see in incredible detail. By scanning with both eyes, so the viewing area of the two boomerangs comes together and forms an X-shape, the spider can see the details of what it is viewing. Two of the spider's other six eyes are located beside the main eyes and the other four eyes are behind them. These are much simpler than the main eyes, but are excellent motion detectors, giving the spider a 360-degree view of its world. The eyes either side of the main eyes are essential for guiding the spider to look at the detail of an object. Researchers have described this as similar to looking around a dark room with a flashlight, while motion detectors tell you where to look.

SNIFFING PERFUMES

Native to East Africa, *Evarcha culicivora* (Salticidae) dines on mosquitoes filled with human blood. The spiders are able to target blood-engorged female mosquitoes and drain them of their blood meal. Male mosquitoes do not drink blood.

BLOOD-BASED PERFUME

Once the blood is inside the spider it is not only used as a meal. It is also used to make a perfume (pheromone) that makes both males and females more attractive to each other. The blood-based aphrodisiac lowers the threshold of both sexes to mating. Possibly, the bloody perfume tells both sexes that they are good at hunting and would make good parents for their offspring.

Given spiders are mostly solitary, many different adult females produce airborne perfumes to tell males where they are located and that they have not mated. By following the female's scent the male can find her. He can then court her to see if she considers him a worthy suitor to father her offspring.

↓ A female mosquito (*Anopheles gambiae*) with an engorged abdomen after feeding on blood, possibly passing on malaria.

↓ When a female wolf spider *Pardosa milvina* (Lycosidae) detects a courting male's silk, she spins more of her own attractive silk.

→ Sitting on a leaf near Lake Victoria, in Kenya, a female vampire spider, *Evarcha culicivora*, feeds by sucking human blood from the body of a female mosquito that has recently fed on a person.

BUILDING AN ORB WEB

Researchers in the United States wanted to see the details of orb web building to gain insights into how a spider's tiny nervous system can build such an elaborate structure.

ARTIFICIAL INTELLIGENCE

Using six female hackled orb web spiders—*Uloborus diversus* (Uloboridae), native to the western United States—researchers at Johns Hopkins University built an arena in their lab to record their web building, which takes place at night over a number of hours. They used infra-red cameras and lights, so they could film the spiders in the dark. Recording every movement the spiders made during web building would have involved analyzing millions of individual movements. But the researchers had help from an artificial intelligence program that had been trained to track the movement of limbs. It monitored the movements of 26 different points on each spider—3 points on each of its 8 legs and the front and rear of the spider—as the 6 spiders built 21 different webs.

STICKY STRETCHY DROPLETS

Orb web spiders have very elastic webs that can absorb the impact of a flying insect without breaking. Gluey droplets are also added to the spiral lines of orb webs, which look like beads arranged along a string. The droplets are not only sticky but, like the web, also very elastic. When prey fly into the web, the droplets stretch but do not detach from the silken threads. As the web relaxes from the impact, the stretched glue also bounces back to the shape of droplets. This causes them to stick to prey and trap it in the web, before the spider rushes across and captures its detained meal.

↑ The orb web stabilimentum of a species of *Cyclosa* (Araneidae) from the Philippines may deter birds from flying into it. *Cyclosa* are known as trashline orbweavers.

~ Web development ~

The researchers found that each spider built its webs using similar movements for each stage of the webs' construction. The finished webs may have looked slightly different from one another, but the steps required for their construction were very predictable. Just knowing the position of a spider's legs told researchers what part of the web the spider was building.

~ Tiny brains ~

The results suggest to researchers that a web-building program is encoded into spiders' brains. It is amazing that such a high level of behavioral complexity can come from such a tiny brain.

LIQUID GOLD

The evolution of silk in spiders has been compared to the evolution of flight in insects as being responsible for the incredible diversity of species in both groups. The strength and elasticity of silk has seen it evolve for use in a seemingly endless number of ways, from ingenious prey-trapping devices to silken retreats high in the Himalayas and ones beneath the sea. While many people associate spider silk with webs, new discoveries, such as the lassoing acrobatic ant-slaying spider, see it used in almost unimaginably clever ways.

Spider silk is stored as a liquid in silk glands inside the spider's abdomen. There are ten different types of silk glands and spiders can have up to seven different glands, with each gland producing different types of silk. For example, web builders have glands that produce silk for building webs, capturing prey, wrapping prey and eggs, ballooning, and making safety draglines.

WEB DECORATIONS

Some of the spiders that build orb webs decorate the hub of their web with a silken decoration known as a stabilimentum. These may be a spiral or cross of thick lines, or some other shape. The St. Andrew's cross spider, *Argiope keyserlingi* (Araneidae), is named after the cross-shaped stabilimentum in its web. These silk decorations were so named because it was originally thought they helped stabilize the web. Research has shown this is not the case. The decorations are more likely used to deter predators like birds from flying into the web. Since the only orb web spiders with stabilimenta are in their webs during the day, it seems a good explanation for reducing risk for an exposed web spider.

SPITTING SILK

Spitting spiders (Scytodidae) produce silken threads from silk glands in their abdomen. They also have a large gland in their head filled with a silk-like liquid and glue. They use this mixture as a liquid weapon to trap prey, similar to the sticky threads of an orb web. The mixture is fired out of the spider's enlarged fang ducts in long, thin fibrils. Since the spider shakes its head from side to side as it fires, the prey is covered in two zigzag lines of liquid silk and glue. Almost immediately the silk shrinks by about 60 percent and this contraction and the sticky glue trap prey long enough for the spider to inject its venom.

FROM LIQUID TO SOLID

Pulling silk as it emerges from the spider's spinnerets (silk-spinning organs at the end of the abdomen) causes a change in how the proteins making up the silk are arranged, so they form a solid thread. The liquid silk travels along a narrowing duct until it reaches spigots, which are tiny, tube-like pores on the spinnerets, and from here it is pulled out as a silken thread.

FROM LIQUID TO LIQUID

The glue droplets used to trap prey, mostly flying insects, in orb webs remain as a liquid from inside the silk gland to when they are applied to the spiral lines of the web. Once on the web the elastic sticky droplets stop prey from escaping.

↓ *Argiope picta* (Araneidae), a species of orb web spider, pulls out swathes of silk and uses this to wrap prey it has just captured.

HANGING BY A THREAD

J ust as mountain and rock climbers often attach themselves to an anchored length of rope to prevent injury if they fall, many spiders are also anchored as they move around their environment. In the case of jumping spiders, which can jump up to around forty times their own body length, an anchored safety line of silk means that if they fall, they only drop the length of the lifeline. The safety line also stops the spider's body from drifting to the left or right as it jumps and is used as a brake to slow the spider down as it lands. This means the spider lands on its legs and is not traveling as fast as when it first jumped.

FLEEING AND FEEDING

Web spiders will drop on a dragline of silk when threatened by predators or parasitoids. A crab spider—species of *Mystaria* (Thomisidae)—from Africa hangs from a silk line while feeding.

↓ Having jumped onto a leaf, a female goldenrod spider, *Misumena vatia* (Thomisidae), still has a dragline of silk.

↓ The jumping spider *Hypoblemum albovittatum* (Saliticidae) climbs a dragline and gathers up the silk.

→ This female zebra jumping spider, *Salticus scenicus* (Salticidae), is preparing to launch herself from a leaf. After attaching a safety dragline of silk to the leaf, she increases the fluid pressure in her rear legs. They straighten and she becomes airborne.

PROTECTIVE COCOONS

F emale spiders lay eggs and use silk to varying degrees to protect the developing embryos once they are outside her body. Some species only cover the eggs with a few strands of silk and guard the eggs inside a silken nest. But many species build elaborate silken cocoons with a soft, pillow-like layer on which the eggs lie and a tough, thick outer layer of silk to keep the humidity inside the cocoon high enough to prevent the eggs drying out.

SILKEN FORTRESSES

The silken fortresses created by some spiders also protect the eggs from egg parasitoids, like some flies and wasps, which will lay their own egg inside a spider egg, as well as predators that eat spider eggs. Often, egg cocoons are camouflaged with debris, while others match the color of the background, so they blend in. Spiderlings emerge from the eggs and remain inside the cocoon until they have molted at least once before leaving.

↓ Young cobalt blue tarantulas, *Cyriopagopus lividus* (Theraphosidae), emerge from an opening in their egg sac.

↓ Female ant-mimicking jumping spiders *Myrmaplata plataleoides* (Salticidae) guard their eggs inside a silken nest.

→ With a bowl-shaped egg sac, a female *Argiope bruennichi* (Araneidae), sits in her web. The baby spiders remain inside the egg sac during winter and then emerge in the summer. The female's yellow markings act as a lure to flying insects.

SPIDERS CAN FLY

Since the 19th century it has been observed that very small spiders can travel vast distances by using silk to become airborne. The spider points its abdomen upward and releases a thread or threads of silk. Wind currents lift the spider off the ground and into the air. Called ballooning, although it is more like kiting, it was thought that this is how spiders came to colonize remote islands. Traveling by air for spiders can be very hazardous, unless they only travel a short distance and can alight on land. Longer flights increase the risk of being preyed on by birds or landing in the sea. Spiders have even been picked up by planes that were flying at high altitudes.

ELECTRICALLY CHARGED SPIDERS

Spiders have also been observed kiting when there was no wind to lift them. Spiders weighing around a hundred times more than was thought possible for kiting have also been seen flying. Recently, it has been shown that the threads of silk released from spiders before they become airborne are negatively charged. Given that the surface of the Earth is also negatively charged, the electrostatic repulsion between the negatively charged silk and the Earth allows the spiders to become airborne without any air currents.

~ Sensing electricity ~

Special sensory hairs on spider legs, called trichobothria, can detect electric fields. In a large, windless container researchers showed that spiders ballooned when an electric field was present. They did not balloon when the field was turned off.

→ Sitting on a tiny seedhead, this money spider—a species of *Tenuiphantes* (Linyphiidae)—prepares to be carried away, attached to its dragline, by wind currents.

FLYING IN THE 19TH CENTURY

In 1883, an enormous volcanic explosion blew apart the island of Krakatoa, located off the coast of Java. Three months after the explosion, visitors to the new island created from the explosion found small spiders living there. The only way they could have arrived on the island was by kiting from the mainland 20 miles (32 km) away.

When Charles Darwin (1809–1882) was traveling the world on HMS *Beagle*, he noticed hundreds of tiny spiders had drifted onboard the ship when it was 60 miles (100 km) from the coast of South America. They had also ballooned from the mainland and then ballooned from the *Beagle* and continued on their journey.

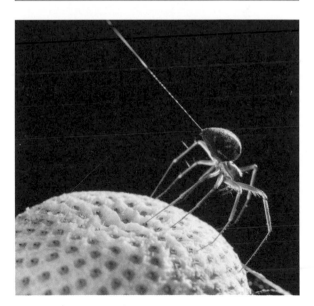

ANT EATERS

L ess than 1 percent of known spiders eat ants because they are aggressive and can sting and bite. However, some spiders have evolved very novel ways to use silk, so the ants' defenses do not pose a threat.

ACROBATIC ANT SLAYER

The Australian ant-slayer spider, *Euryopis umbilicata* (Theridiidae), is probably one of the most remarkable ant-eating spiders. This spider can even take on an aggressive ant, the banded sugar ant, which is twice its size. It also captures its prey around 90 percent of time. This is almost double the success rate of a predator like a lion or wolf. During the day the spider hides under the bark of eucalyptus trees. After dusk the ant slayer emerges and, facing downward, attaches itself to the bark with a single silken line. It waits for an ant to walk by as it climbs the trees looking for food.

WRAPPING AND RUNNING

Flatmesh weaver spiders—species of *Oecobius* (Oecobiidae)— build flat webs, which are used for shelter and waiting for prey rather than hunting. These spiders are also known as star-legged spiders because all their legs point outward like an eight-legged starburst. This leg arrangement helps the spider with its unusual method for catching ants, its common prey. Attached to the spider's shelter are long lines of silk that act as trip-wires if prey comes into contact with them. When the spider picks up vibrations from a trip-wire, it rushes from its shelter and runs around the prey, covering it in a shroud of fine silk strands. It then bites the prey and takes it back to its shelter to feed.

ANT BASKET

The Australian basket web spider, *Saccodomus formivorous* (Thomisidae), is not only unusual for what it eats but also how it traps its prey. This spider eats ants and catches them with a basket that looks like a lobster pot. While the pot has a diameter of about $7/16$ in (11 mm) and a length of about $9/16$ in (14 mm), it is a rigid structure, unlike a web. The spider taps its legs on the basket to lure ants inside, then quickly injects them with venom. Researchers have found that the basket always contains an egg sac made of a similar silk to the one used by spiders to wrap their eggs. The basket web spider has used this silk not only to protect its eggs, but also to make an elaborate trap for ants.

~ Somersaulting spider ~

Using high-speed cameras, researchers filmed what an ant-slayer spider does when it launches an attack. It happens so quickly that it is impossible to see without slowing down the action. When the ant is close enough, the spider launches itself into the air and, traveling at about 10 in (25 cm) a second, somersaults over the ant, dragging a line of sticky silk behind it. It attaches this line to the ant before hanging freely while still attached to the silk. These acrobatics take place in about a tenth of a second. The spider then climbs back toward the ant and covers it in more sticky silk before injecting it with venom and dragging it away for feeding. The acrobatic ant slayer could be the inspiration for a spider-like superhero!

→ A green weaver ant (*Oecophylla smaragdina*) in an aggressive pose: holding open its very sharp mandibles.

SPERM WEBS AND EGGS

Spiders have a pair of short, leg-like appendages called palps on either side of the chelicerae. These are used by males and females as sensory organs and to help hold prey. In males they are also modified to hold and transfer sperm to inside the female's epigynum (genital opening).

CHARGING PALPS

Although males make sperm in their abdomen, there is no internal route for the sperm to travel from this site to the palps. Instead, males make a silken sperm web that is used to temporarily hold the secreted sperm before it is drawn up inside the palps. Some male spiders have palps that look similar to those of females, while others resemble miniature boxing gloves. Once males have molted a final time and are

adults, they transfer sperm to their palps. With charged palps the behavior of the male spider changes: it may no longer feed and instead will go searching for females. While most females live for a year or two, many adult males are only around for a few weeks.

FERTILIZING EGGS

Females produce eggs in their abdomen and various ducts connect the eggs to the female's genital opening. During mating the male's palp fits inside the female like a lock and key. Most males can only transfer sperm to females of the same species.

STORING SPERM

The sperm is stored inside the female in special packages called spermathecae and when she is ready to lay eggs, the stored sperm is used to fertilize the eggs.

← Male jumping spider *Cosmophasis umbratica* (Salticidae) from Singapore courts a nearby female by showing off his bright yellow pedipalps and dance moves.

COURTING WITH AND WITHOUT GIFTS

Spider courtship can be complicated since they are predators and it is prey rather than potential mates with which they usually interact. While most female spiders are bigger than the males, both sexes will cannibalize each other. Males and females need to be on guard and tread carefully in the mating game.

Since different spiders perceive the world with different senses, courtship signals are sent in many different ways and are usually initiated by males. Most web builders do not have good eyesight and males court by plucking the web to signal their intentions to females. Jumping spiders have excellent eyesight and males will perform elaborate dances for the females. Wolf spiders have reasonable eyesight and the males wave their legs at females and drum their palps on the ground. Females often prefer the fastest drummers. Displays identify the male as a potential mate rather than a meal, though both outcomes are possible.

COLORFUL DANCER

Probably the most colorful spider courtship displays are by tiny ($1/16$–$1/4$ in/2–6 mm) jumping spiders from Australia—around 50 species of *Maratus* (Salticidae). They are called peacock spiders and it is easy to see why. While the females are plain brown, the male's third pair of legs are elongated and tipped with white tufts of hair. Many males have fan-like flaps on their abdomen, so they can make it wider and turn it into a courting canvas covered in scales that produce a kaleidoscope of garish colors. By waving their abdomen and legs in synchrony and dancing energetically, the males try to convince females they would make great fathers.

In Australia, a female jumping spider—a species of *Euryattus* from Queensland (Salticidae)—makes her home inside a rolled-up leaf hung from silken threads. Male *Euryattus* climb down the silk and court the female by shaking her leaf with vibratory signals. If she emerges and is keen, he will dance for her. *Portia fimbriata* (Salticidae) is a jumping spider that hunts other spiders. The male climbs down onto the female's leaf and mimics the vibratory courtship display of male *Euryattus* to entice the female out of her leaf, so he can eat her.

Pheromones on the silk of the North American female wolf spider *Gladicosa bellamyi* (Lycosidae) can tell the male if he should attempt to court her. They tell him whether she has mated or not before and, most importantly, whether she has cannibalized previous suitors. It is not surprising that males avoid courting female cannibals.

GIVING GIFTS

Whether they are strummers, drummers, or dancers, males also show their worth as a suitor by giving the female the insect equivalent of a box of chocolates—the perfect gift might be an insect wrapped in silk.

~ Deceptive gifts ~

Some males nibble on part of the gift and give the female a partially eaten box of insect chocolates. Other males that couldn't quite capture an insect gift give the female a silk-wrapped leaf. The next level of deception is to give the female an empty silk wrapping. By the time the female realizes she has been duped with the empty wrapping, the male has mated and left.

↓ A male nursery web spider, *Pisaura mirabilis* (Pisauridae), carries a worthless gift in an attempt to deceive a courting female.

MALES FIGHTING MALES

When engaged in combat, two males will eye each other up before locking tusks and pushing and shoving. While it's easy to imagine this is a fight between elephants, it is actually between two tusk-bearing jumping spiders.

TUSKED SPIDERS

With a body length of around ³/₁₆ in (5 mm), the tiny jumping spider *Thorelliola ensifera* (Salticidae) is common in rainforests in Southeast Asia. Just below the two large, forward-facing eyes, which are common to all jumping spiders, the males have two curved tusks (around 0.02 in/0.5 mm long), which extend horizontally out from the spider's face.

These are enlarged hairs, which are also found on females, but they are much smaller and look like the hairs covering the spider's body. Another difference is that the chelicerae of females are convex (curved outward), while those of males are concave (curved inward).

~ Contests of strength ~

When the little tuskers eye each other up, the smaller male usually runs away. It is only when the two males are the same size that they come together and lock tusks for a few seconds. This usually ends with one male unlocking their tusks and running away.

~ Charging males ~

Sometimes, if the contest of strength is not decided quickly, the fight escalates and the males will charge at each other, locking tusks and jaws as well as shoving each other with their outstretched front legs. A full-on charge is settled after one male retreats and runs away. The victor is likely to be a successful suitor for nearby females.

← The forward-facing curved tusks of the jumping spider *Thorelliola ensifera* are used in contests of strength with rival males. Only males of the same size compete in this way.

FEMALES FIGURING OUT MALES

Like the females of most species, female spiders invest a lot more in reproduction than males. Most female spiders are bigger than most males and need to eat more to have sufficient nutrient reserves to produce eggs. After laying eggs, females vary in how much they invest in protecting and feeding their offspring. All that males invest in the development of offspring is mating with females that use their sperm to fertilize their eggs. It is not surprising that many females want an indicator of the genetic fitness of the male that will father their offspring.

DANCING MALES

Females may choose males on the basis of their elaborate courtship behavior. Many male jumping spiders have garish colors and impressive dance moves. This is to show females that they are a good investment and, by sharing their genes in the next generation, they will produce males with impressive dance moves and females that make good choices.

← Waving his front pair of legs, the jumping spider *Habronattus pyrrithrix* (Salticidae) courts a female.

← With extended legs, a male spotted wolf spider, *Pardosa amentata* (Lycosidae), courts the female by waving his palps.

→ While the rather drab-colored, but camouflaged, female peacock jumping spider, *Maratus speciosus* (Salticidae), looks on, a colorful male performs elaborate dance moves with his legs and body. His colorful abdomen is fringed with orange hairs, which are only visible during courtship.

MATING WITH CANNIBALS

Mating with cannibals is how the reproductive lives of spiders is often imagined. The most infamous spider cannibal is probably the black widow spider *Latrodectus mactans* (Theridiidae), although the females only kill and rarely eat the males after mating. The myth may have started because males and females were often kept in cages from which the males could not escape and so they became prey. In nature, males almost always live to mate another day.

FEMALE CANNIBALS

Female redback spiders from Australia, *Latrodectus hasselti* (Theridiidae), cannibalize over 60 percent of males during mating. The female has two genital openings leading to separate sperm storage organs. During mating, the male inserts his palp into one of the openings and transfers sperm. He also twists the rear end of his abdomen into the fangs of the female who begins to feed on the male. The male is much smaller than the female, so he is more of a snack than a full meal. To guarantee he will father all or most of her offspring, the male also needs to insert his second palp into the female's second genital opening.

GRUESOME SUCCESS

For one tiny male orb web spider, which weighs only 1 percent of the female, mating is violent and fatal, but it has nothing to do with the behavior of the female. Before courting, the male self-amputates and discards one of his two sperm-carrying palps. When the male finds a female willing to mate, he inserts his other palp into the female's epigynum and dies. His heart stops, but the hydraulic pressure in his body focuses on transferring sperm to the female. Then she dines on the tiny snack.

DEADLY SILENCE

For some male six-spotted fishing spiders (*Dolomedes triton*; Pisauridae) in North America, courting mated females leads to a silence that speaks volumes. Doomed males that detect a nearby female start courting by sending out vibratory signals across the water's surface. When there is no reply to their signals, they move closer to the female, expecting a courting reply. Soon the silent female pounces, grabs, and eats the male whose courting signals have led to a less-than-desired outcome.

Of course, the male can only do this if he is still alive. However, the odds are not looking good because he is being consumed and still needs to transfer sperm from his second palp. If he fails to transfer sperm from his second palp, then a second male who mates with the female will transfer his own sperm to the female's second sperm storage organ. This means the first male will only father half of her offspring. But males do something extraordinary with their abdomens while courting females. They contract muscles and constrict their abdomens, so the part of their abdomen being eaten by the female is sealed off from the rest of the body. This buys the male enough time to transfer sperm from the second palp to the female before he dies and is eaten. All rather gruesome, but it does mean the male will be the father of the female's offspring. In closely related species of *Latrodectus*, where the risk of sexual cannibalism is low, males do not constrict their abdomens.

↓ Mating with the highly cannibalistic female redback spider, *Latrodectus hasselti*, can be a life-changing experience for males.

MATERNAL INSTINCTS

Many female spiders show no maternal care beyond laying their eggs in a protective egg sac. Others stay to protect and feed their offspring until they can survive on their own. Some male spiders allow themselves to be eaten by the female. After mating, they become a nutritious gift to help the development of the offspring they have fathered. Similarly, some female spiders become the last meal their offspring have before leaving the nest.

FEEDING BABIES

A social Australian crab spider *Australomisidia ergandros* (Thomisidae) lays only one clutch of fertilized eggs. The female keeps a second clutch of unfertilized eggs inside her body, which she converts to hemolymph and her babies start to feed on her body (behavior known as matriphagy). Over a few weeks, her body shrinks as the babies grow bigger. Larger cannibalized females produce bigger and more offspring since there is less sibling cannibalism than those feeding from a smaller female.

↓ A female fishing spider *Pisaurina mira* (Pisauridae) protects her egg sac by holding it with her chelicerae.

↓ Offspring of the desert spider *Stegodyphus lineatus* (Eresidae) gather around their mother's mouth to feed on regurgitated food.

→ A female wolf spider (Lycosidae) with around 100 babies on top of her abdomen. The female carries her egg sac around and when it hatches the babies climb onto her back, clinging to each other and to hairs on her abdomen.

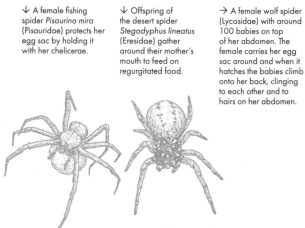

SPIDER SYRINGES

Almost all spiders produce venom, which they inject through a pair of fangs into their prey to paralyze them. Two venom glands each have a duct that travels from the gland and down to the fang duct. When muscles around the gland contract, venom is squeezed out of the gland and moves down and out of the fang duct. Spiders can alter how much venom they inject into prey by controlling how much the gland is squeezed. If the initial injection does not subdue the prey, the spider will inject more venom.

HOW FANGS STAB

The fangs are located at the bottom of the muscularly powerful basal segments of the chelicerae, which help the spider stab prey. Within the basal segments are muscles that open and close the fangs.

TARANTULA VENOM GLANDS

In the majority of spiders, the venom glands are partially located in the basal segments of the chelicerae and extend out into the cephalothorax. Surprisingly, large tarantulas have relatively small venom glands that fit entirely within the basal segment of the chelicerae.

SPITTING SPIDER GLANDS

Spitting spiders have enormous venom glands, which can fill most of the cephalothorax. The glands have two lobes, with the front part containing venom and the rear lobe containing glue and liquid silk (see *Spitting Silk*, page 49).

← A threatened
male trapdoor spider
Cantuaria dendyi
(Idiopidae) from New
Zealand rears up, raises
his palps, and reveals
his large fangs.

A PAINFUL EXPERIMENT

In November 1933, Dr. Allan Walker Blair, a doctor at the University of Alabama medical facility, allowed himself to be bitten on the finger by a female black widow spider. He was cynical that a small spider could cause the severe symptoms reported by patients. After two hours of writing down increasingly alarming symptoms, the unfortunate doctor could no longer write and his assistant took over. Soon he was taken to hospital and over the next few days suffered a range of symptoms. He was worried he would go insane and was given morphine.

CURED

The doctor who treated him said, "I do not recall having seen more abject pain manifested in any other medical or surgical condition." Dr. Blair concluded in a report on his experiment: "The venom injected by the bite of the adult female spider, *Latrodectus mactans*, is dangerously poisonous for man." The doctor and his cynicism were both cured.

↓ Being bitten on the finger by a female black widow spider, *Latrodectus mactans* (Theridiidae), is not a good idea.

↓ The face of a female black widow, showing her fangs.

→ The red hourglass marking on the underside of the abdomen of female black widow spiders, *L. mactans* (Theridiidae), has been shown to act as a warning to birds not to attack. It is less obvious to insect prey.

STOPPED IN THEIR TRACKS

Spiders are the main predators of insects and they have evolved an arsenal of different strategies to capture and immobilize them. Part of their success is due to the evolution of silk and venom. While a variety of different webs are used to trap prey, it is the use of venom that has enabled spiders to paralyze or kill prey that is their size or much bigger, including occasionally animals like birds, bats, and snakes.

COMPLEX VENOMS

Almost all spider venoms are a complex mixture of different chemicals, including neurotoxins, which work together to interfere with the prey's nervous and consequently muscular systems.

POTENT *PORTIA* VENOM

Species of *Portia* are jumping spiders that specialize in preying on other spiders, including other jumping spiders. They have evolved a venom that seems particularly fast acting when it is injected into another spider. Since *Portia* is hunting another predator that could attack and kill *Portia*, it makes sense that the prey is subdued quickly. If *Portia* injects venom into another spider about its size, the prey staggers a short distance and becomes paralyzed in around 10 to 30 seconds. Prey about twice the size of *Portia* take about 15 to 30 minutes to succumb to the venom, before the spider will approach its large prey.

~ Neurotoxins ~

Most spiders have between 10 and 30 different neurotoxins because they capture a variety of different insect prey. A neurotoxin that paralyzes one type of insect may not be effective on other prey. So spiders have evolved lots of different tools in their venom toolbox to allow for a varied diet. After venom has been squeezed out of the venom glands and injected into the prey's body, neurotoxins in the venom can prevent nerves from sending signals to muscles and prevent muscles from contracting, effectively paralyzing the prey and preventing it from escaping.

~ Enzymes in venom ~

Another component of spider venom are enzymes. These are different from the enzymes secreted from the midgut for digestion. The enzymes in venom break down some tissues, including connective tissue, which allows the venom to penetrate farther into the prey's body beyond the site where it was first injected.

← Sitting on a daisy, a goldenrod spider, *Misumena vatia* (Thomisidae), eats a bee, caught when it landed on the flower.

PEOPLE AND VENOMS

S piders and snakes are among the animals most feared by people. It is estimated that around 4 percent of people have arachnophobia, where their fear far outweighs the real risk spiders pose and this can interfere with their daily lives. Even in places without any dangerous spiders, a fear or dislike of spiders persists. Fortunately, for people with arachnophobia, many different treatments exist.

BITING SPIDERS

It is estimated that globally around five people die each year after being bitten by a spider. In comparison, around 60,000–100,000 people die each year from snake bites, with the majority of these in the Indian subcontinent. Although Brazilian wandering spiders (species of *Phoneutria*) and Australian funnel web spiders (Atracidae) can be aggressive, most spiders do not bite people. Why would they? We are not prey to them and they avoid large, warm mammals. The most likely scenario in which a spider will bite a person is when they are accidentally squashed against our body and have fangs large enough

---

CELLAR SPIDER MYTH

There are many myths about spiders, with one of the most common involving the widely distributed and common cellar spider, *Pholcus phalangioides*, also known as the daddy-long-legs spider. The myth suggests that the venom of this spider would be deadly to people, if only its very small fangs could penetrate our skin and it could inject its deadly cocktail. Studies of the venom have shown that it is not dangerous to people. The myth may have started because the cellar spider is known to catch and eat other spiders, including the venomous black widow spider. People have assumed its venom must be especially potent for us if it can kill a black widow, although this turns out to be an urban myth.

FUNNEL WEB SPIDER

There are around 35 species of Australian funnel web spiders and six species are known to produce venom that can lead to life-threatening symptoms, if the spider injects it into a person. The venom is neurotoxic and can paralyze the victim, leaving them unable to breathe. Around 12 deaths from the venom of funnel web spiders were recorded over a hundred years, until the early 1980s. Since 1981 an anti-venom has been available and there have been no recorded deaths from this time. While the venom can be dangerous to us and other primates, it does not affect cats, dogs, and rabbits. Male funnel webs have the most dangerous venom and can be aggressive during the mating season.

to penetrate our skin. The most dangerous non-human animal on Earth is the mosquito and mosquito-borne diseases kill around 725,000 people a year. Unlike spiders, mosquitoes deliberately seek out and bite people.

~ Dangerous spiders ~

While many spiders can bite and produce a variety of non-lethal symptoms, of the more than estimated 60,000 different species of spiders only a few have venom that could be lethal to people. In the United States, the bite of the female black widow spider, *Latrodectus mactans*, has been fatal in the past, but because of an anti-venom there have been no deaths since 1983. Similarly, an anti-venom is available for the potentially dangerous bite from Brazilian wandering spiders and Australian funnel web spiders.

→ The development of spider anti-venoms has meant the chance of dying from a spider bite is very unlikely.

UNUSUAL PREY

While spiders are usually thought of as predators that use venom to capture insects and other spiders, they also catch a number of unexpected prey. Probably the most unusual and largest invertebrate (animals without backbones) prey seen being eaten by a spider was a 39 in (1 m) earthworm. It was captured by one of the largest tarantulas, the Goliath bird-eating tarantula, *Theraphosa blondi* (Theraphosidae), which is known for eating earthworms far more often than birds. This tarantula may also hold the spider record for catching and eating the largest vertebrate (animals with backbones), a cane toad, with a body length around three times that of the spider.

There are many records of spiders catching snakes, frogs, and lizards. However, the stickiness of their webs and the potency of the venom of widow spiders allow them to catch snakes (see below) as well as small mammals like mice and rats. Spiders with large orb webs, like various species of *Nephila*, have been seen catching small birds and bats. The prey becomes tangled in the web's sticky threads before being injected with venom.

JUMPING SPIDER
EATS FROGS AND LIZARDS

Phidippus regius (Salticidae) is one of the largest jumping spiders and the females can have a body length, including legs, of up to 7/8 in (22 mm). Like all jumping spiders, this species has excellent eyesight and is found in the southeastern United States, where it is especially common in Florida. While most jumping spiders feed on insects and other spiders, this species has added some very unlikely items to its menu. Researchers have observed the spider catching anole lizards and tree frogs. Most spiders are opportunistic feeders and this species appears big enough to take on unusual prey that most jumping spiders would not be able to capture.

BLACK WIDOW SPIDER
EATS SNAKES

The black widow spider, *Latrodectus mactans*, is probably the most well-known venomous spider in the United States because its venom can be dangerous to people. But the same venom is also very toxic to another group of vertebrates: snakes. Black widow spiders and other widow spiders belong to a group of spiders (Theridiidae) commonly called tangle-web spiders. Their webs are very strong and have sticky threads attached to the ground. Trapped snakes are injected with venom before being hauled up into the web and eaten. With a strong, sticky web and powerful venom, widow spiders can catch snakes up to thirty times their own size.

→ A lizard trapped in the web of redback spider, *Latrodectus hasselti* (Theridiidae), will soon become prey.

FISHING IN FISHPONDS

In Sydney, Australia, a fishing spider was seen dragging a goldfish with a body length of $3^{1}/_{2}$ in (9 cm) out of a garden fishpond. The spider must have been many times smaller than the captured goldfish. Similarly, in South Africa, a fishing spider was observed dragging a pet goldfish named Cleo that was over twice its size from a garden fishpond. *Dolomedes triton* is a fishing spider common in North America. While this spider is well known for catching fish from lakes and streams, it has also been reported dragging mosquito fish over twice its length from garden ponds.

FLESH-DISSOLVING SPIDER VENOM

The black widow spider, *Latrodectus mactans* (Theridiidae), and the brown recluse spider, *Loxosceles reclusa* (Sicariidae), are probably the most feared spiders in the United States, though bites from both species are rare. The venom of the brown recluse contains enzymes that can cause necrotic skin lesions. As its common name suggests, this spider is reclusive and rarely bites people. A more likely culprit for the skin lesions thought to be caused by the brown recluse is an antibiotic-resistant bacterial infection (methicillin-resistant *Staphylococcus aureus* or MRSA). Misdiagnosis is common, even in regions where the spider is not found.

BROWN RECLUSE AT HOME

In 2001, over 2,000 brown recluse spiders were found living in a house in Kansas, which they shared with four people for many years. The people living in the house did not know they were brown recluse spiders and often saw them. Nobody in the house was ever bitten.

↓ The brown recluse spider, *Loxosceles reclusa*, is also known as the violin spider because of the violin-like markings on the cephalothorax.

↓ This brown recluse spider has caught an armyworm (*Spodoptera frugiperda*).

→ Hoping to be chosen as a mate, a male brown recluse spider, *L. reclusa*, approaches a female, attracted by airborne chemical cues she has deposited on her silk. When he approaches the female he performs vibratory displays to avoid becoming prey.

BOWL AND DOILY WEB

T he most familiar spider webs are vertical, cartwheel-shaped orb webs, but some spiders build horizontal sheet webs, and one of the most unusual is very common in Central and North America. The bowl and doily spider, *Frontinella pyramitela* (Linyphiidae), is named because part of its web looks like a bowl and appears to hover above a flat sheet web that resembles an ornamental lace mat called a doily.

TUMBLING PREY

The bowl is suspended by many hanging threads and the doily is held in place by silk attached to vegetation. The web is not sticky, but small flying insects that hit the hanging threads tumble into the bowl. The spider hangs upside down underneath the bowl, then pulls its victims through the silk and injects them with venom. The doily part of the web is thought to act as silken barrier to predators that could otherwise easily attack the tiny spider.

↓ When prey, such as the fungus gnat (right), hits the web made by the bowl and doily spider, *Frontinella pyramitela* (left), it tumbles into the bowl and the waiting spider.

→ In the background, a female bowl and doily spider *F. pyramitela*, adds silk to her horizontal sheet web (doily) below the main inverted bowl-shaped web. The female in the foreground is looking for a suitable site to build her web.

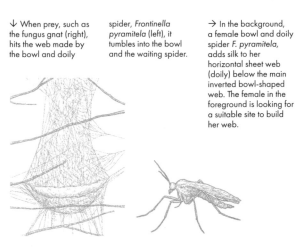

WALL OF WEBS

I n the National Botanic Gardens, in Entebbe, Uganda, close to Lake Victoria, hand-sized female golden silk orb-weavers (species of *Nephila*; Araneidae), with webs around 3 ft (1 m) in diameter, trap numerous flying insects. They are known as golden silk orb-weavers because the very strong silk used in their webs is golden colored. What makes this group of spiders and their webs different is the fact that around 50 of them have built connected webs that have become a wall of orb webs measuring about 30 ft (10 m), the height of a two-story building. Since the lake nearby provides the nutrients for a huge number of flying insects, the spiders are never short of a meal because their enormous aerial filter catches prey.

As well as grabbing insects from their own webs, golden silk orb-weavers also travel to nearby webs after picking up the silk-guided vibrations of trapped prey. Although golden silk orb-weavers are usually solitary predators and aggressive to intruders, they tolerate neighbors wandering into their webs since there is no need to compete for food.

CAST OF MILLIONS

In the 19th century, a species of golden silk orb-weaver spider, *Trichonephila inaurata* (Araneidae) from Madagascar, was milked for its silk to make fabric for weaving. In the 21st century, two textile experts, eighty spider wranglers, and around two million golden silk orb-weavers from the highlands of Madagascar were involved in a project to use 19th-century techniques to extract spider silk in order to produce fabric for weaving different garments. Since the silk from around 23,000 spiders is needed to produce about 1 oz (28 g) of silk, it was a project requiring a cast of millions.

TINY MALES

Mature male *Nephila* are around 10 percent the weight of mature females, which can weigh over 1/5 oz (6 g), heavier than many hummingbirds. Being tiny compared to the enormous females carries risks for many males. When the males mature, they cannot spin sticky threads and so steal prey from the webs of females. Some end up, not as a worthy suitor but as a snack for the female.

↓ A tiny male approaches a female *Trichonephila inaurata* on her web, hoping he will be a worthy suitor.

~ Missing limbs ~

A number of males will compete for the same female on her web and many lose limbs due to close encounters with cannibalistic females. A strategy males use to reduce the chances of being eaten by the female is to mate while she is distracted by wrapping or feeding on prey.

A SMELLY MOTH TRAP

Many moths can escape from orb webs. If they fly into a web, small scales on their bodies and wings detach and adhere to sticky droplets on the silk strands of the webs, and they escape. Little bald patches on the moth's body serve as testimony to how it survived flying into an orb web.

BOLAS SPIDERS

One group of orb web spiders (family Araneidae) have evolved an ingenious way to trap moths. And all it takes is a few strands of silk. Bolas spiders, especially species in the genus *Mastophora*, are common in North America. A bolas spider attaches a strand of silk between two bits of vegetation, which it holds onto with its rear legs. With one of its front legs it holds a line of silk, which has a sticky droplet (or bolas) at the end. To set the trap, the spider also makes itself attractive to moths.

DECEIVING MOTHS

The yellow garden spider, *Argiope aurantia* (Araneidae), is named because it is common in gardens in North America and builds a large orb web. It is also known to build its web close to a light source, like a porch light, to catch flying insects attracted to the light. One entomologist noticed that a spider on his porch had caught a number of male oakworm moths, which only fly during the day. He watched as they flew from some distance straight into the web. Apparently, the spider produces a pheromone that tricks the moth into behaving as if they are flying toward a receptive female, rather than a devious spider—day and night hunting using two very different strategies.

↑ A bolas spider,
Mastophora cornigera
(Araneidae), preparing
to hunt moths by
supporting itself in a
web with its rear legs.

~ Moth perfume ~

The spider emits a pheromone that mimics the perfume produced by female moths to attract males. If a male moth starts to follow the scent, the spider will pick up the vibrations of its fluttering wings as it gets closer. It then starts to twirl the silken bolas by rotating its front leg.

~ Sticky end ~

Not only is the bolas very elastic, but the center is also much stickier than the outside. This means that if the moth hits the bolas, the less sticky outside can slide past the detachable scales and the more sticky center sticks to the moth; a sticky end for an amorous male lured by a fragrant and fateful perfume.

TRAP DOORS

Trapdoor spiders live in silk-lined burrows, capped with a silken hinged trapdoor. The trapdoor is often decorated with bits of vegetation found close to the burrow, making it almost impossible to see that it is the opening to a spider's home. Females spend their entire lives within the burrow.

TRAPPING PREY

At night, these nocturnal hunters partially open the trapdoor and extend their two front pairs of legs out of the burrow. They keep their rear legs inside the burrow, so they cannot be accidentally locked out of their home. Sensory hairs on the spider's legs pick up the vibrations of insects, often ants and beetles, walking nearby. When the prey is close enough to be grabbed, the spider flings open the trapdoor, scoops up the insect, and pulls it down into the burrow, where it is eaten. Prey remains are periodically taken out of the burrow and left as grim little piles nearby.

↓ The lidless burrow of the Brisbane tube spider, *Arbanitis longipes* (Idiopidae), is often decorated with pieces of vegetation.

↓ Female Brisbane tube spiders spend most of their lives inside their silk-lined burrows.

→ With a hinged lid disguised to match the surrounding vegetation, the home of the South American trapdoor spider, *Actinopus pusillus* Actinopodidae), is very well camouflaged. At night, the female opens the lid, sits at her burrow's entrance, and waits for prey to appear.

SLINGSHOT SPIDERS

The triangle weaver spider, *Hyptiotes cavatus* (Uloboridae), is native to the United States and Canada. It belongs to a group of spiders without venom glands and builds a triangular orb web.

WOOLLY WEB

The web does not have sticky silk strands like an orb web. Instead, it has woolly silk strands that entangle prey. When the spider has finished making the web, it does something no other spider is known to do—it turns the web into a slingshot.

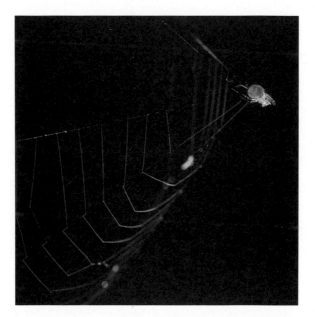

CATAPULTING AFTER MATING

Philoponella prominens (Uloboridae) is a species of hackled orb web spider common in Asia. Females almost always cannibalize males after mating unless the male can escape. Males have evolved a very novel way to avoid becoming victims of sexual cannibalism. After mating, the male pushes against the female's body with his first pair of legs. This causes a joint in the lower part of each leg to bend the leg. By suddenly forcing fluid into these legs, they straighten and the male becomes airborne and is catapulted to safety. This propulsive feat happens in about 4 milliseconds and the male travels at a speed of around 35 in (90 cm) per second.

~ Making a slingslot ~

To make a slingshot, the spider holds the front edge of the web with its front legs while its rear legs start winding up the silk attached to nearby vegetation. This dramatically increases the tension the web is under, like a bow being pulled taut.

~ Using a slingslot ~

When prey hits the web, the spider releases the spring-loaded silk strand and both it and the web are propelled forward 1 in (2.5 cm) at a speed of around 2,300 ft (700 m) a second. That is sixty times faster than a cheetah can accelerate. No muscle-powered mechanism could ever move that quickly—it is only a spring-loaded web that could trap prey in this way. The acceleration of the web and its movement as the tension is relaxed entangle the prey, without the spider having to make contact with it. If the prey is not sufficiently tangled for the spider to start feeding, it can reload the web and release it again, until its meal is safe to eat.

← The triangle weaver spider, *Hyptiotes cavatus*, uses its web as a slingshot, so that it can quickly catapult itself toward its prey.

CASTING A NET

I t has a pair of eyes two-thousand times better than our own at seeing in the dark. It uses webs to catch prey, which it throws at them. During the day it hides in plain sight, but it is so cryptic, researchers say it is almost impossible to detect. The ninja-like net-casting spider—a species of *Deinopis* (Deinopidae)—has the biggest eyes of any spider. It looks as if it is wearing a pair of round sunglasses that are far too big for its face. To some, this gives the spider a fierce and sinister look and it is, in fact, also called the ogre-faced spider. Its eyes are even more sensitive in the dark than owl or cat eyes.

HUNTING IN DARKNESS

At night, the net-caster builds a frame shaped like the letter A and suspends itself on a silken anchor from this frame. By facing downward it can use its enormous eyes to watch the vegetation below. With its two pairs of front legs it holds a net made of woolly strands of silk. Its spidery night-vision googles can pick up the movement of prey like beetles as they walk. Some net-casting spiders even drop their white spider excrement onto the vegetation below to make prey stand out as they walk across it. When the spider picks up six legs walking, it stretches its two front legs, which also stretches the net and makes

---

CATCHING FLYING INSECTS

While staring at the ground in search of prey, the net-casting spider is also "listening" for the sound of flying insects above. Its legs have sensory hairs that are sensitive to the sound made by the flapping wings of insects like moths and mosquitoes. These hairs can also detect how close the insect is to the spider. When the insect is close enough to be caught, the spider makes a remarkable acrobatic movement. In about a fifteenth of a second, it moves its front legs and the net above its head. It then does a backward flip and catches the insect as it flies into the net.

SPIDER WEARS A BLINDFOLD

Researchers at the University of Nebraska-Lincoln wanted to see how well *Deinopis spinosa*, from Florida, could hunt if it was wearing a temporary blindfold and could not use its incredible eyesight to see prey moving in near darkness. The blindfold was made from an opaque dental silicon and applied to the spider's large eyes with a wooden toothpick. When wearing the blindfold, the spider took ten times longer to catch a cricket. This demonstrated the evolution of the spider's large eyes was essential for successfully hunting in near darkness, so it could avoid diurnal predators.

it larger. Then at an incredible speed it drops down with the net and scoops up the prey. It jiggles the prey in the net to entangle it further, then uses its rear legs to wrap it in a silk parcel before injecting the struggling insect with venom.

STICK MIMIC BY DAY

After a night of hunting, and as dawn approaches, the net-casting spider changes from formidable predator to harmless stick. It stretches its two pairs of front legs in front of its thin body and its two pairs of rear legs behind it. This makes the spider invisible to possible predators like birds, which would easily spot the net-caster if it looked like a spider.

~ Rebuilding eyes ~

If the spider anticipates that it will be in light rather than darkness, it breaks down the light-sensitive membrane in its big eyes. This may be so the eyes can cope with being in ambient light. After all, the spider cannot close its eyes! At dusk it rebuilds the membrane so it can see what other spiders are unable to see in the dark.

↓ The face of the well-named ogre-faced spider (species of *Deinopis*) with its enormous eyes, the biggest of any spider.

PORTIA: A SPIDER-EATING SPIDER

E very spider's nightmare is *Portia*, a genus of jumping spider found in the tropics. It specializes in hunting other spiders, has eyesight to rival that of a primate, and the cunning of a carnivorous mammal.

CRYPTIC HUNTER

This spider resembles a piece of dead leaf and walks like a baby robot learning how to walk. Both these characters help disguise this skillful hunter, which has an array of different strategies to trick and trap its eight-legged prey.

~ Hunting web spiders ~

The main prey of *Portia* are web-building spiders. *Portia* plucks the spider's web to make the resident curious, rather than behaving like an insect is struggling in the web—the predator does not want to become prey. When the web spider is close enough to attack, *Portia* grabs it and injects venom. *Portia* can walk on the silk of other spiders' webs and may even feed in a trapped spider's web.

LEAPING AND HUNTING

Like all jumping spiders, species of *Portia* have excellent eyesight. *Portia* also specialize in hunting and eating other spiders (araneophagy), including jumping spiders. In Australia, *Portia fimbriata* (Salticidae) has a very clever way of catching its common jumping spider prey, *Jacksonoides queenslandicus*. Using chemical cues from its prey, *Portia* can be aware the spider is close by, even before it is seen. When it detects that prey is near, it leaps randomly into the air. This may make the prey spider reveal itself by moving and looking at the leaper, as if it was possible insect prey. By keeping one leap ahead, *Portia* can then focus on stalking and catching the duped spider.

↑ Hunting other spiders is dangerous, but it is a specialty of *Portia*, a jumping spider that even hunts other jumping spiders. This spider is known for its ingenious hunting methods.

If *Portia* cannot easily pluck the spider's web, it makes a detour out of sight of its prey and launches an attack from above. Web spiders have poor eyesight and do not see *Portia* descend on a silken line. Once *Portia* is close enough to the web spider, it grabs it.

~ Hunting spitting spiders ~

Portia even takes on spitting spiders, a formidable prey. With its keen eyesight *Portia* looks for a female spitting spider holding a clutch of eggs with her fangs. This means she cannot spit and is safe to attack.

FISHING SPIDERS

The six-spotted fishing spider, *Dolomedes triton* (Pisauridae), from North America, catches small fish when they swim close to the water's surface. The spider rests its two front legs on the water to "listen" for movements below, while its two rear legs attach to a solid surface. This is similar to how a web spider picks up vibrations from its web. If a fish swims by, the spider dives underwater and catches the fish before dragging it onto dry land and eating it. These spiders have even been known to catch goldfish from fishponds in people's gardens! When the spider goes underwater, it traps an air bubble so it can breathe. It also hides underwater to escape predators. Since the air bubble makes the spider buoyant it needs to cling to something under the water, or it will pop back up to the surface.

FISHING IN A PITCHER PLANT

Another diving and fishing spider spends its life inside a plant with a very small "swimming pool." Pitcher plants look like jugs and are partially filled with fluid to drown insects. The pitcher produces nectar to attract insects to its slippery surfaces. The insects often fall

---

BREATHING UNDER WATER

Nepenthicola misumenops can hide from predators, like parasitic wasps, in the fluid inside the pitcher plant by trapping a bubble of air over its book lungs (breathing organs) on the underside of its abdomen and breathing underwater for up to 40 minutes. The spider also uses its "scuba tank" to go fishing for insect larvae living in the fluid. The pile of dead insects at the bottom of the pitcher is called the necromass and it is home to mosquito larvae. The spider enters the necromass, agitates the water, and flushes out the larvae, so it can catch them. It then drags the larvae out of the fluid onto the dry pitcher wall and feeds.

DIVING BELL SPIDER

The diving bell spider, *Argyroneta aquatica* (Dictynidae), is found in Europe and Asia, often in lakes and ponds. It is the only known spider to spend almost its entire life under water. Like other spiders that are able to breathe underwater, the diving bell spider can hold a bubble of air over its abdomen and book lungs. To do this, the spider builds a canopy of silk attached to pieces of vegetation and then fills it with air bubbles that it brings down from the surface. This provides enough air for the spider to go for long periods of time without having to return to the surface. From its aquatic home, the spider can grab aquatic insects, crustaceans, and even small fish as they swim by.

to a watery grave, so the plant can extract nutrients from their bodies. *Nepenthicola misumenops* (Thomisidae) is a crab spider from Asia and uses the pitcher as a home, a source of food, and a nursery for its offspring.

~ Catching ants ~

When it comes to catching ants, which are far too dangerous to catch alive, the crab spider has help from the pitcher plant. The spider will pick up vibrations from an ant that has fallen into the fluid. Once the ant has sunk and drowned, the spider waits about ten minutes before retrieving the harmless fresh prey.

→ Like many other species of *Dolomedes*, the fishing spider *Dolomedes aquaticus* (Pisauridae), from New Zealand, also catches small fish as prey.

SPITTING SPIDERS

Spitting spiders (species of *Scytodes*) have large heads that are armed and dangerous (see *Spitting Silk*, page 49). When the spider is about ³/₈ in (10 mm) from its prey, it measures how far away it is by carefully probing with a front leg, since, like most spiders, it does not have good eyesight. Then it fires a sticky mixture.

GLUING DOWN PREY

When spitting, the spiders move their fangs rapidly from side to side, so prey are covered in a zigzag pattern of glue, which contracts further, entangling and trapping the victim. The glue is fired at a speed of about 98 ft (30 m) a second in three-hundredths of a second. The amount of spit spat depends on prey size. Venom is injected to further subdue the sticky meal. They really are an eight-legged spitting cobra.

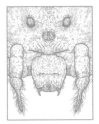

← The face of a spitting spider (a species of *Scytodes*), showing its small fangs, which inject venom and squirt sticky glue when hunting prey.

↙ A female spitting spider (a species of *Scytodes*) uses her fangs to hold her egg sac.

→ Having just spat liquid silk and glue, the spitting spider *Scytodes thoracica* (Scytodidae) has trapped a long-jawed orb web spider (species of *Leucauge*). When the spit hits the target, it dries and shrinks, further ensuring there is no escape for the victim.

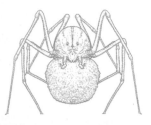

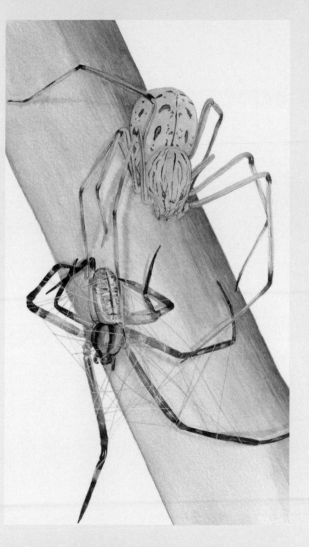

AS STILL AS A STATUE

Crab spiders are so named because they can move sideways like a crab. They are known as sit-and-wait or ambush predators. Many species sit motionless on vegetation or flowers with their long front legs stretched out, waiting to embrace insects that land or walk nearby. While most crab spiders are cryptic, and blend in with their background, some can even change color depending on the color of the flower they are sitting on. Changing from white to yellow to pink seems to be more about fooling predators than deceiving insects visiting flowers, since they see color very differently to a bird looking for a spider snack.

INTACT PREY

While it be may of little consolation to an insect that becomes a meal, crab spiders leave their prey almost totally intact after feeding. But all the insides are removed and these are extracted through a couple of sets of fang holes, usually one set in the head and the other in the abdomen. When the spider grabs an insect it injects venom through the body part closest to it fangs. Once the insect has stopped struggling, the spider almost always positions the prey so it can start

---

DRINKING NECTAR

An unusual source of food for some crab spiders is not found inside prey, but rather the nectaries of flowers. In North America, male *Misumenoides formosipes* (Thomisidae) can weigh twenty times less than a female and sixty times less than a female full of eggs. Mature males are only around for a few weeks and do not feed on prey, focusing instead on finding females and mating. While females get their mass from prey, tiny males will drink nectar from flowers, which lengthens their life span. If dehydrated, they will visit up to 80 flowers an hour to replenish lost fluids.

---

feeding from the head, which contains nervous tissue full of fats and proteins. With large prey the spider may even rotate the head with its legs, so it is sealed off from the thorax and abdomen.

~ Empty prey ~

The crab spider then inserts its fangs into the head to create a couple of holes through which it can suck out the tissues inside. It creates a seal over the holes and uses its powerful sucking stomach to create a vacuum in the prey's head. When it relaxes, the spider releases digestive enzymes, which are drawn into the head as the vacuum equalizes and which begin digesting tissue. The spider repeats a cycle of sucking and relaxing as it mixes enzymes with tissue, until it finally retains the partially digested meal and completes digestion in its own digestion system. The spider not only uses the prey as a source of food, but also as an extension of its own digestive system as it cycles fluid back and forth.

→ A bird-dropping crab spider, *Phrynarachne ceylonica* (Thomisidae), has captured and is eating a blue bottle fly.

WOLF SPIDERS

Wolf spiders are named because some species run after prey, just like a wolf, although they do not hunt in packs. Some wolf spiders can walk on water. They hunt insects as well as dive underwater to catch tadpoles and small fish.

RUNNING WITH OFFSPRING

Some female wolf spiders carry their ball-shaped egg sac by attaching it to the spinnerets at the rear of their abdomen. The spinnerets are slightly raised, so the egg sac does not drag on the ground. This is very helpful, as the female can continue to hunt while carrying the egg sac. It must be a bumpy ride for the developing baby spiders as their mother runs and hunts. After hatching, the baby spiders climb up their mother's legs and cling to hairs on her abdomen and to each other, riding a spider rollercoaster for the next week or two before they leave home.

← The face of a wolf spider *Hogna radiata* (Lycosidae) showing its two large eyes, which give it reasonably good eyesight.

↓ A female wolf spider with her egg sac, which she carries around with her, even when hunting.

→ Hunting like a wolf and a wolf spider, a running crab spider—species of *Thanatus* (Thomisidae)—has captured a jumping spider—species of *Habronattus* (Salticidae). It is likely the crab spider picked up vibrations through the vegetation, as the jumping spider walked.

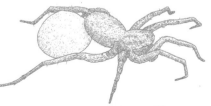

A VAMPIRE-KILLING SPIDER

A jumping spider from East Africa, *Evarcha culicivora* (Salticidae) has a taste for human blood. The spider does not target our blood directly, but instead looks for female mosquitoes that are very good at accessing human blood, often without us being aware of it.

A TASTE FOR BLOOD

Like all jumping spiders, *Evarcha culicivora* has very good eyesight, so it can recognize the red, blood-filled abdomen of a female mosquito against a background of very similar looking flies. These spiders can also smell blood and when they scent a blood-filled female, they are primed to look for her and her blood meal.

WATCHING MOVIES

Professor Robert Jackson and Dr. Fiona Cross, of the University of Canterbury, in New Zealand, have even projected movies onto a small screen to see what type of prey *Evarcha culcivora* jumps on. It almost always chooses a female mosquito that has been feeding on human blood. Male mosquitoes do not drink blood. Drinking blood from inside mosquitoes has its advantages, as all spiders are fluid feeders and break down their prey's tissues with digestive enzymes—the blood is like fast food for the spider.

~ Blood as fast food ~

When *Evarcha culicivora* ruptures the mosquito's abdomen, a mostly digested liquid lunch awaits it—rather like drinking soup from a bowl. Another spin-off for these blood-drinking spiders is that they turn the blood into perfume that makes both sexes more attractive to each other. Having a taste for miniature vampires certainly pays off in more ways than one for this vampire slayer.

FEEDING FRENZY

Evarcha culicivora detects the odor of a human-blood-filled mosquito and starts searching for its next meal. Researchers were interested to see what the smell of blood would do to the spider if it could see a number of blood-filled female mosquitoes at the same time. The spider was put in an arena with a number of mosquitoes and the odor of human blood was wafted in. The spider went into a feeding frenzy and killed up to 20 mosquitoes. This was far more mosquitoes than the spider could eat and most were discarded. It seems when the spider smells blood and sees lots of prey, a killing spree follows.

↑ Sucking human blood from a mosquito (*Anopheles gambiae*) turns the body of the jumping spider *Evarcha culicivora* blood red.

SAVED BY A
SPIDER SCULPTURE

Bits of dried vegetation often become stuck to spider webs. For trashline orb web spiders—species of *Cyclosa* (Araneidae)—these are the building blocks for cunning disguises that camouflage the spider and hide it from predators.

HIDING AMONG TRASH

Combining the debris with eaten insect remains and molted exoskeletons, tiny trashline orb web spiders create a line of trash down the center of their webs. By blending in with the garbage the spider can hide during the day when predators such as birds scan webs for a juicy spider. One species of *Cyclosa* in Taiwan uses its web debris to make life-sized replicas of itself. In that way, a predator is more likely to get a mouthful of rubbish than the resident spider when it attacks.

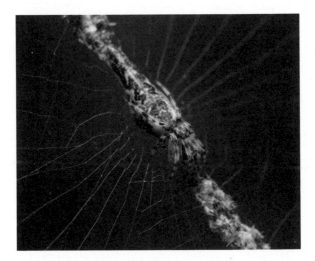

BRANCH STUMP MIMIC

The common name, tree stump orb weaver, for various species of *Poltys* (Araneidae) perfectly describes what type of spider these are and what they look like. These spiders have a brown abdomen shaped like a stumpy branch. During the day, the female hides among vegetation with her abdomen raised and looks nothing like a spider. At night she builds an orb web and at dawn hides with her clever disguise. Recently, in southwest China, a species of *Poltys* has been discovered where the females mimic a leaf. The upper side of their abdomen looks like a live green leaf, while the underside resembles a dead leaf. Their abdomen also tapers to look like a stalk detached from a stem.

GIANT SPIDER DECOY

The most remarkable web decoy is built by different species of *Cyclosa*, one from Peru and one from the Philippines. They build a large replica of a spider, complete with eight legs. It is thought this is to fool a predator into believing that the spider in the web is too large to attack.

~ Spider puppeteer ~

These spiders were first discovered by scientists working in Peruvian rainforest who thought they saw a large dead spider in a web. It then started to move because it was being shaken by the spider puppeteer above the decoy. While the spider in the Philippines builds a spider sculpture with its eight legs spread out, the one in Peru builds a sculpture with its legs pointing downward. It's not only people who build scarecrows to deceive birds. Some spiders build scary spider sculptures.

← A trashline orb weaver (species of *Cyclosa*) hides in its web, among the debris made mostly from the remains of its prey.

STAYING OFF THE MENU

Jumping spiders (Salticidae) belong in the largest spider family, which contains about 5,000 species. In the genus *Myrmaplata* they all mimic ants. Since jumping spiders have excellent eyesight and hunt during the day, they are vulnerable to predators like birds.

Many predators that attack spiders will not attack ants because of their aggressive group defenses. While ants do not welcome spiders into their colonies, living close by and looking like an ant may be enough to keep the spider in "ant clothing" safe. One of the most extreme forms of ant mimicry is seen in *Myrmaplata plataleiodes*, which mimics the green weaver ant (*Oecophylla smaragdina*). The spider walks on six legs and waves its front pair of legs like ant antennae. It has dark eyespots to mimic ant eyes and walks like an ant following a chemical trail.

A REMARKABLE TRANSFORMATION

Until the males' final molt to maturity, they look just like females. But during the final molt males undergo a remarkable transformation. Within the safety of a silken nest the male's jaws change from pointing downward to horizontal. They are then extended and enlarged until they are 50 to 70 percent the length of the spider's body. The spider's fangs also extend to the same length as the jaws. With the jaws extended horizontally, the fangs are neatly tucked underneath them.

← Looking remarkably like a leaf is an orb web spider (species of *Poltys*) recently discovered in forests in China.

MALES MIMIC TWO ANTS

An obvious question is whether having exaggerated mouthparts compromises the appearance of male *Myrmaplata plataleiodes* spiders as green weaver ant mimics. This is not the case, however, because there are two sizes of weaver ant workers. The larger ones forage for food while the smaller ones care for eggs and larvae inside the nest. Larger workers often carry smaller workers to different parts of the nest and the smaller workers look just like the spider's enlarged mouthparts. A species of *Myrmaplata* in Sri Lanka even has dark eyespots on the ends of its jaws to further the illusion that the male ant mimic is just a larger worker carrying a smaller worker.

The male spiders equip themselves in this way so they can compete with other males for nearby female mates. When two males approach each other, they each unsheathe their long, sword-like fangs and wrestle with their fangs and jaws. It is a contest of strength and sometimes the fights escalate, with the loser being thrown up into the air. While the winner stays put, the other spider runs away.

SWORDS WITHOUT VENOM

Since the male has such long fangs, there is no continuous duct from the spider's venom glands to the duct at the end of the fangs. As a result, adult males cannot inject venom into prey to paralyze them. Adult females, however, can still inject venom into prey since their fangs are not modified.

~ Stabbing prey ~

Males have to stab their prey, which is far less effective than injecting venom. Large prey often escape and the body parts from smaller prey are sometimes skewered on the male's fangs. All in the name of love!

SAVED BY BIRD DROPPINGS

Daytime hunting web spiders and ambushing crab spiders sit and wait for prey and can be targeted by predators like birds. If the spider mimics something unattractive to the bird, it is likely to be left alone. What could be better than looking like a bird dropping that has fallen from the sky and landed on a web or a leaf?

MIMICKING DROPPINGS

Spiders that mimic bird droppings tend to have shiny bodies, which adds to the illusion that the droppings are wet and fresh. Some spiders even sit on silk spun into a shape that looks like a fresh bird dropping sitting on a dried dropping. The bird dung crab spider, *Phrynarachne ceylonica* (Thomisidae), from Southeast Asia adds another layer to its dropping doppelanger. It looks and smells like bird dung. This added layer of deception keeps away birds and attracts prey like flies.

↓ An orb web spider (species of *Cyrtarachne*) mimics a bird dropping to fool predators.

↓ Another species of orb web spider, *Celaenia atkinsoni* (Araneidae), also looks like a bird dropping.

→ *Phrynarachne katoi*, the bird-dropping crab spider, is a diurnal sit-and-wait predator. Not only does it have a shiny body that looks like a fresh bird dropping, but it also sits on silk spun into a bird-dropping shape.

III

SPINNING OUT OF CONTROL

T he greatest enemies of spiders are parasitoid wasps. Many inject venom into spiders to paralyze them, so they become a harmless larder for the wasp's offspring. By eating the spider the offspring can develop from an egg to a larva before pupating and emerging as an adult. The spider's body is their pantry and sometimes nursery.

LARVA FEEDS ON SPIDER

Once the egg of *Reclinervellus nielseni* (Ichneumonidae) has hatched, the larva partially extracts itself from the egg sac and makes small holes into the spider's abdomen, so it can feed on its hemolymph. It is thought the larva mixes the blood with an anticoagulant to stop it clotting. After a day or two, the larva leaves the egg sac completely and uses special hooks to latch onto the spider's abdomen. It continues to feed and grow for two to three weeks. During this time the spider keeps behaving like a spider and catching prey on its web, even when the larva grows to about a third of the size of its abdomen.

← A parasitoid wasp larva (*Reclinervellus nielseni*) feeds on a still-living orb web spider (species of *Cyclosa*).

WASP TRICKS SPIDER

Some parasitoid wasps can manipulate the behavior of spiders in insidious ways, and one lives in the jungles of Costa Rica. The orb web spider *Leucauge argyra* (Tetragnathidae) builds a typical-looking orb web to trap prey. The ichneumonid female wasp (*Hymenoepimecis argyraphaga*) targets the spider as a host for its offspring. The wasp uses two different methods to attack the spider. It may hover next to its web, before grabbing the spider and injecting it with venom to paralyze it. Its second method is more devious. The wasp behaves as if it is prey trapped in the spider's web. When the spider comes close to the wasp, it is attacked and paralyzed.

First, the wasp checks whether another female has already laid an egg on the paralyzed spider. If the female finds another egg or larva, it kills this before attaching her own egg to the spider's abdomen. She then flies off. The wasp's venom wears off after five to ten minutes, and the spider carries on unawares.

~ Spider becomes zombie ~

Once the larva is ready to pupate, it injects a chemical into the spider. This makes the spider drastically change its web-building behavior. The manipulation of the spider's nervous system causes it to spin an initial sequence of web building in an endless loop that results in a structure that look nothing like an orb web. Instead, it resembles thick lines of silk, like the spokes of a wheel with most of the spokes missing. Soon after the zombie spider finishes its master's retreat, it dies as the larva sucks it dry.

~ Web protects cocoon ~

The larva tosses the spider's corpse to the ground before spinning its own cocoon in the middle of the lines of thick silk. Suspending the cocoon above the ground provides a safer place to pupate away from possible predators. About a week later, the adult wasp emerges and flies off, so its offspring can control the mind of another spider.

FROM BURROW TO DEADLY BATH

In New Zealand, some female trapdoor spiders live for over 20 years and remain in their silk-lined burrows covered with a protective trapdoor all their lives. Females that leave their burrows are likely to have been infected by a nematode worm that has been growing inside the spider for many years.

AQUATIC LIFE AND DEATH

The nematode starts life underwater inside an insect larva that pupates and moves onto land carrying the parasite. If the infected insect is eaten by a trapdoor spider, the nematode takes up residence inside the spider. It grows larger and larger in the spider until it causes the female to change her behavior, leave her burrow, and seek water, where she drowns. The worm, now the length of a piece of spaghetti, emerges from the spider and mates. The offspring then seek aquatic larvae and the life cycle starts again.

↓ Tiny parasitic mites cling to the cephalothorax of a female trapdoor spider *Cantuaria dendyi* (Idiopidae).

↓ Fruiting bodies of a parasitic fungus (species of *Gibellula*) emerge from the body of a jumping spider.

→ Although the female trapdoor spider *Cantuaria dendyi* spends her entire life inside a burrow, the parasitic nematode (*Aranimermis giganteus*) causes the spider to leave her burrow to find water. When she drowns, the 12 in (30 cm) worm can emerge.

A TARANTULA'S
WORST NIGHTMARE

Most spiders use venom to paralyze or kill their prey before feeding. Parasitoid wasps use venom to paralyze, but not kill, different spiders. They want to keep their victim alive for as long as possible for nightmarish reasons.

TERRIFYING WASPS

The most terrifying wasps are probably some species of pompilids. Tarantula hawk wasps are the biggest spider parasitoids and attack a group that contains the largest spiders: tarantulas. Although the adult wasps only feed on nectar, their offspring feed on spider flesh.

↓ A tarantula hawk wasp has paralyzed a tarantula with venom, leaving it is alive but harmless. The spider's body will now become food for the hawk wasp's offspring.

DEATH BY FLY

The Acroceridae, commonly called small-headed flies, are parasitoids of various spiders, including tarantulas. The larvae find a tarantula host by searching or waiting for a spider to walk by. A larva then enters the spider's body through the soft membranes between leg joints, or directly into the abdomen. Once inside, the larva makes its way to the spider's book lungs, so it can breathe the same air as the spider. The larva may remain inside the spider for a number of years before finally killing it by eating all its internal tissues. Shortly before the spider dies, it builds a small web. The fat larva emerges from the spider's body and uses the web to pupate to maturity.

~ Paralyzed ~

When a female wasp finds a tarantula, it injects the spider with venom through its long stinger. The wasp then drags the helpless spider to its burrow and lays an egg on its body. After the egg hatches into a larva, it makes a small hole in the spider's abdomen and burrows into the tarantula's body. It feeds for a couple of weeks and by avoiding eating the spider's vital organs, the tarantula remains alive and fresh. Finally, the larva finishes feeding and pupates inside the spider's hollowed-out body before emerging as an adult wasp and leaving the burrow. If the wasp is a female, it will follow in its mother's footsteps, while males look for opportunities to mate with females.

~ Painful sting ~

The sting of the tarantula hawk wasp is not especially dangerous to humans, although it is considered one of the most painful insect stings known. The pain only lasts for about five minutes, but apparently it is five minutes you will never forget.

ESCAPING DANGER

While spiders are formidable predators they are also food for other predators and parasitoids. Some have evolved very inventive and surprising ways of escaping.

ROLLING SPIDER

The huntsman spider *Carparachne aureoflava* (Sparassidae) is found in the Namib desert in Southern Africa. Its common name is golden wheel spider because of its color and the way it escapes from its most dangerous enemy, a parasitoid wasp. While the spider will try to hide in the sand, if the wasp comes too close, it does a short run before folding its legs into semicircles, turning sideways, and then, with the help of gravity, rolling down a slope like a runaway wheel. By making up to 44 turns a second, it can escape from the wasp.

SPIDER GYMNAST

In a southeastern desert in Morocco there is a spider that behaves like an eight-legged version of a human gymnast. The flic-flac spider, *Cebrennus rechenbergi* (Sparassidae), is a huntsman spider that escapes from predators by performing some remarkable moves. It briefly runs, then propels itself into the air using its rear legs and lands on its front legs. It keeps moving in this way until it is safe.

FASTEST RUNNER

While rolling and tumbling are very unusual ways for spiders to escape danger, running is not, and many spiders flee just by running away. The record for the fastest runner is the giant house spider, *Eratigena atrica* (Agelendiae), found in North America and Europe. While this spider builds a messy, trampoline-like web, it will run after prey and away from predators. With long, slim legs and a slim body it can cover 20 in (50 cm) in a second.

→ When the flic-flac spider, *Cebrennus rechenbergi*, performs spider gymnastics to escape, it moves twice as fast as when it walks.

PREY MIMICS PREDATOR

The markings of some insects, including flies, bugs, and moths, mimic the legs and face of jumping spiders. Making the spider believe it is looking at another jumping spider, rather than prey, may give the insects a chance to escape. Metalmark moths in the genus *Brenthia* from Costa Rica raise their forewings and hindwings to reveal markings that mimic the legs and face of a jumping spider. They also move in a similarly jerky fashion to a jumping spider. Dr. Jadranka Rota, while at the University of Connecticut, found that the jumping spider *Phiale formosa* (Salticidae), from the same area as the moths, was deceived by them. These moths were more likely to survive than a similar sized moth with plain wings.

SPIDER INFLUENCERS

For people who do not like spiders, the coordinated choreography of eight legs walking and often moving in unpredictable directions is frequently mentioned as contributing to their fear. But the same mechanics of spider locomotion that induce fear in people have also been an inspiration for the design of robotic legs. Spiders use hydraulic leg extension by forcing hemolymph into their legs and coordinating this with muscle flexion of the leg joints. Similar mechanisms have been incorporated into robotic leg design to produce legs for a range of different robots, depending on the robot's use.

SPIDER ROBOTS FOR EXPLORING MARS

The golden wheel spider, *Carparachne aureoflava* (Sparassidae), of the Namib desert in Southern Africa has developed an ingenious method of evading a dangerous parasitic wasp (see page 118). Its escape strategy has inspired the design of a robot that could be used to explore the

---

VENOM WITH A SILVER LINING

Australian funnel web spiders include species with venom that can be lethal to people. Before the development of an anti-venom, it could kill a person in 15 minutes. The venom contains an arsenal of different neurotoxins and other molecules that evolved to paralyze insect prey. Researchers in Australia have discovered that one of the neurotoxins in the funnel web spider *Hadronyche infensa* (Atracidae), instead of being dangerous to us, may reduce the severity of brain damage following a stroke, where part of the brain is deprived of oxygen. This deprivation can lead to a cascade of brain cell death, but the neurotoxin from the venom turns off a vital neural pathway and prevents this from happening.

difficult terrain on Mars. Called the Spider Rolling Robot, the designers see it as an alternative to wheeled explorers that resemble a highly modified car. They also think the robot might be energetically efficient because it could be assisted by the windy conditions on Mars and, just like the spider, use gravity to roll down hills.

SPIDER ACROBATICS INSPIRE ROBOTS

The flic-flac spider, *Cebrennus rechenbergi* (Sparassidae), from Morocco's Erg Chebbi desert (see *Spider Gymnast*, page 118), has been mimicked in a robot that can walk or roll on six legs. When it rolls, six legs act like two wheels and two steer it.

NECROBOTICS

Engineers in the United States recently developed a type of robot using the body of a dead wolf spider as inspiration. One of the engineers noticed that the legs of dead spiders curl inward. This is because the spider no longer has a pressurized hydraulic system to extend its multi-jointed legs. The engineers inserted a syringe into a spider's rigid front cephalothorax. By changing the amount of air pumped into and out of the spider, it could extend and close its legs, as if alive. The corpse could pick up objects such as another dead spider. They named this new field of robotics necrobotics, and hope to use it as a model for developing small mechanical grippers. A bizarre and unexpected use of a dead spider.

← A flic-flac spider, *Cebrennus rechenbergi*, has just landed on its front legs, after launching itself using its rear legs.

TOOL-USING SPIDER

In the windswept Namib desert of Southern Africa a tube-dwelling spider has evolved a novel way to detect vibrations from passing prey. The corolla spider—species of *Ariadna* (Segestriidae)—lives in a silk-lined burrow. It surrounds the entrance with small pebbles. These are almost always made of quartz crystal and the spider usually uses seven pebbles. Each pebble is attached to the burrow entrance with a small line of silk.

GOOD VIBRATIONS

The spider sits at the entrance of the burrow with its six front legs in contact with the silk. Quartz crystals are very good at transmitting vibrations. If a passing ant or beetle happens to touch or walk over the crystals, the prey's vibrations travel through the pebble and the silk line to the spider. It knows exactly where the vibrations are coming from in its stone circle, and will rush out and pull the prey back into the burrow.

← The carpenter ant (*Componotus maculatus*) is common prey for the tool-using spider (species of *Ariadna*).

↙ A tool-using spider (species of *Ariadna*) walks around the desert looking for small pebbles to build its elaborate trap.

→ A female corolla tube-dwelling spider (species of *Ariadna*) is resting her legs on quartz crystal pebbles surrounding her burrow entrance. She is waiting for vibrations from prey as it touches the pebbles, which alert her when a meal is nearby. A carpenter ant is about to reveal its presence.

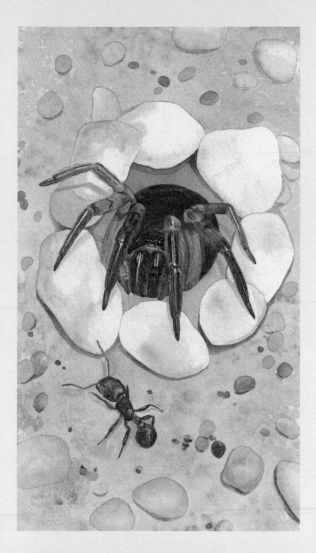

CUTTING-EDGE SPIDER RESEARCH

Advances in technology have allowed researchers to look into jumping spider eyes with a spider-sized ophthalmoscope, or use a video camera with a frame rate almost 1,800 times faster than the standard frame rate to film a spider's snapping jaws. Technology has helped to peel back the layers of the evolutionary history of spiders and reveal their spectacular abilities that at times seem beyond our imagination.

LEAPING JUMPING SPIDERS

Researchers used a high-speed video camera, scanning electron microscopy, and various tools to measure the strength of silk to understand how the common jumping spider *Salticus scenicus* (Salticidae) uses dragline silk when it jumps. Although, like all spiders, jumping spiders use dragline silk as a safety line, it is not known what is involved when a jumping spider jumps and trails a safety line of silk behind it. The researchers filmed the spider jumping across a 1¼ in (3 cm) gap. The spider took off at a speed of about 28 in (70 cm) a second, with a line of dragline silk attached to the takeoff point.

TRACKING JUMPING SPIDER EYES

An ophthalmoscope is a familiar instrument used to look into the backs of our eyes. Researchers in the United States and New Zealand have designed and built the equivalent for peering into the two main eyes of jumping spiders. In spite of their size, jumping spiders have eyesight to rival that of a mammal. When the spider is in position the researchers can show it videos at the same time as using a camera attached to the eye tracker to record what it is looking at. If the spider sees a cricket, the eyes quickly move to focus on what may be prey. Professor Beth Jakob, from the University of Massachusetts Amherst, said: this is a "little window into their mind."

SNAPPING AT HIGH SPEED

Researchers measured the speeds that different species of trap-jaw spiders (Mecysmaucheniidae) can snap their open jaws on prey. The tiny spiders collected from leaf litter in rainforests in New Zealand and southern South America range in size from $1/16$–$1/2$ in (2–10 mm). They can close their open chelicerae in times ranging from 10 milliseconds to a staggeringly fast 0.1 milliseconds, which is about 800 times faster than you can blink. The fastest was a species of *Zearchaea* from New Zealand and this was also the smallest, being the same size as a grain of rice. Using high-speed video cameras required a speed of 40,000 frames a second before the fastest snapper could be seen snapping.

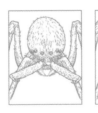

→ The lightning fast snapping jaws of trap-jaw spiders may have evolved to catch springing springtails before they escape.

~ Second toughest known silk ~

Remarkably, the silk was pulled from the jumping spider's body at a rate of 20–28 in (500–700 mm) a second. The researchers also found that the dragline silk is the second toughest silk ever found in a spider, second only to some orb web silk. It would seem this strength allows the silk to be extracted at an incredibly fast rate when the spider jumps at such high speed. The silk also slows the spider down mid-jump and keeps its body orientated in preparation for landing. What happens when a $1/4$ in (5 mm) jumping spider jumps is truly amazing.

LIVING WITH SPIDERS

For many millions of years, spiders have been living almost everywhere on Earth, except Antarctica. Humans have been living almost everywhere on Earth for a very short period. All that time, people have shared living spaces with spiders. They very seldom harm us and most of the time lead shy, retiring lives.

SPIDER SURVEY

A survey of 50 random detached houses of different ages and sizes in Raleigh, North Carolina, in the United States, looked at what arthropods (mostly insects and spiders) were living in each house.

↓ The cellar spider *Pholcus phalangioides* (Pholcidae) is found throughout the world and often lives and builds its web in houses.

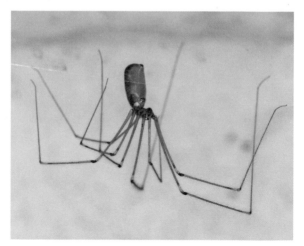

~ At home with spiders ~

Spiders were found living in all the houses, with living rooms, kitchens, and basements being the most frequent spaces shared with spiders. The most common spiders found were cobweb spiders and cellar spiders. One black widow spider was found in a basement. People were very surprised at how many spiders they had as houseguests.

THE BUGS VERSUS THE BRITISH

In 1958, the well-known British spider biologist William S. Bristowe's popular book *The World of Spiders* was published. In it, based on known spider population densities at the time and a conservative estimate that each spider would eat around 100 insects a year, he made a remarkable assessment. He calculated the weight of all insects eaten by spiders in Britain in one year weighed more than the country's total human population (around 52 million in 1958). Given what we know about how much spiders eat globally in a year (see Box) he may have been right.

APPRECIATING THE ELDERLY

A study started in 1974 in a native bushland reserve in southwestern Australia looked at the life history of a native trapdoor spider, *Gaius villosus* (Idiopidae). Females spend their entire lives within a silk-lined burrow and their offspring leave this when they are a few weeks old. The spiderlings build their own burrows close to the one in which they were born. In 1974, the sixteenth spiderling to have its new burrow pegged and monitored was named number 16.

DEATH BY PARASITOID

In May 2016, spider number 16, a female, was still alive, but by October 2016, a parasitoid wasp had made a small hole in the lid of her burrow. The wasp had killed the spider and used it as a source of food for her developing larva. The female was 43 years old when she died, the oldest known age for a spider, and she didn't even die of old age.

↓ After leaving their mother's burrow, the female trapdoor spider *Gaius villosus* digs her own burrow.

↓ The burrow is surrounded by twigs to help the trapdoor spider pick up prey that is walking nearby.

→ The Mexican red-knee tarantula, *Brachypelma smithi* (Theraphosidae), which can live to around 20 years old, is native to Mexico and popular as a pet because of its striking colors and docile nature. Its popularity threatened the existence of this species, and in 1985 it was listed as a CITES protected species.

THE STORY OF ARACHNE

Spiders are woven into the mythical stories of many cultures, often as an allegory concerning the good and bad aspects of human nature. The story of Arachne gave us the first mythical spider and a lesson in humility.

COMPETITIVE WEAVING

Arachne was an excellent weaver from an early age. People would gather to watch her weave. But she became boastful and claimed her skill had nothing to do with the talents and inspiration of Athena, who among other things was the goddess of war and weaving. Athena challenged Arachne to a weaving competition. While Athena wove a tapestry revealing the gods in all their mythical glory, Arachne's tapestry showed their mortal-like flaws. Arachne soon learned this was not a good idea.

Athena was angry that Arachne had woven a tapestry showing the gods in a bad light. She was also angry that Arachne, a mere mortal, was a weaver of talent to rival her, a goddess. She destroyed the tapestry and Arachne left, shameful of her arrogance. Athena eventually took pity on Arachne and decided she could be a weaver forever. There just needed to be a few changes: Arachne grew eight legs and eight eyes, and became the first spider.

ARACHNID LEGACY

Arachnida is the name given to the class of eight-legged animals that includes spiders, harvestmen, scorpions, and mites. It's not surprising that the word is derived from Arachne, the world's first mythical spider, who left us with such a wonderful legacy of fascinating descendants.

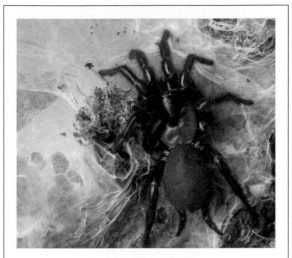

PAINTING ON WEBS

Arachne became the first spider because of her weaving skills and some of the webs of her descendants were used as canvases for paintings. In the 16th century, people started using spider webs, such as those of funnel web spiders (Agelenidae), as a source for layers of silk to create canvases on which they could paint a variety of subjects, from religious figures to landscapes. The collected webs were cleaned of any web debris before being layered on an oval-shaped board, about the size of a postcard. Once there were enough layers, artists would carefully paint with fine-tipped brushes, largely using watercolors with a white pigment base. Since the silk was translucent, the paintings, often kept between glass, seemed to glow in the light.

↑ The silk of the black tunnel-web spider, *Porrhothele antipodiana*, from New Zealand, has been used as a canvas for painting native birds.

ROBERT THE BRUCE

The most well-known story about an inspirational spider is probably that of Robert the Bruce who became King of Scotland. After being defeated six times by the English, Robert was hiding in a cave when he saw a spider trying to connect an anchor thread, so it could build a web. Six times the spider failed to connect the thread to a distant attachment point. On the seventh try the spider succeeded, and this inspired Robert to try to defeat the British one more time and, like the spider, his seventh attempt was successful.

SPIDER FABLE

Historians say the story arose hundreds of years after Robert defeated the English at Bannockburn in 1314 and was, in fact, about a different person. It seems more likely that a family friend, Christina of the Isles, had the resources to supply ships and men, and this contributed to his success. Still, the story of the inspirational spider has persisted.

← In the famous myth of Robert the Bruce, watching a spider changed his destiny.

↙ It is likely Robert the Bruce was inspired to continue his fight against the English by watching the perseverance of a common orb web spider trying to attach its web.

→ When Robert the Bruce was inspired by watching a spider building its web, he may have been watching a female European garden spider, *Araneus diadematus* (Araneidae), construct her orb web. When threatened, the spider vibrates the web violently, so both she and the web become a blur.

TARANTISM

Although its venom is not dangerous to people, medical science in 14th-century Italy reinforced the superstitious myth that the symptoms from the bite of a large Italian wolf spider, *Lycosa tarantula* (Lycosidae), could only be treated with a non-medical cure. The first recorded case of tarantism was in southern Italy in 1370. It was believed its venom caused physical symptoms, including pain, swelling, and nausea, as well as psychological ones like hysteria and "shameless exhibitionism." Victims were often people working in fields during harvest season where the spider was common.

DANCE OF THE SPIDER

As no drugs at the time could cure the patient, it was suggested dancing would flush out the venom and they would be cured. This led to the development of the tarantella dance. Musicians would travel the countryside offering to play music that would inspire the prolonged and vigorous dancing necessary for a cure. Dancing (with breaks) could go on for hours and even days until the patient collapsed and was physically and psychologically free of any symptoms. It was a

---

ITALIAN TARANTULA

The Italian word for tarantula is *tarantola*. While the wolf spider thought to be responsible had tarantula in its name, it's only distantly related to the large group of spiders commonly known as tarantulas. Many tarantula-themed words became associated with this piece of Italian history. The town close to where the first case was described was Taranto, and the effect of the bite was called tarantism. Bite victims were called tarantata. The tarantella, which means "dance of the spider," has persisted through the centuries. While the dance's origin was associated with anguish, today it has morphed into a cheerful dance often performed at weddings.

---

myth that persisted for centuries and spread throughout Italy, reaching a hysterical peak in the mid-1600s. As more medical studies cast doubt on the spider's apparently remarkable venom, tarantism declined and today the tarantella dance is all that remains.

~ Servant guinea pig ~

Oliver Goldsmith, who wrote about natural history, visited Italy in the mid-1700s and saw people dancing the tarantella as a cure for tarantism. He was curious whether the bite of *Lycosa tarantula* really did cause the symptoms for which it had become famous. So he had a spider bite one of his servants. He observed that not much happened other than some local swelling and itchiness around where the spider had bitten the unfortunate servant.

→ The bite of the tarantula wolf spider (*Lycosa tarantula*) was blamed for tarantism, which could only be cured by dancing the long-lasting tarantella.

WEAVING CHARLOTTE'S WEB

The charming book *Charlotte's Web* by E. B. White was first published in 1952 and is probably one of the most popular children's stories ever written. It is the story of how an orb web spider named Charlotte saves the life of Wilbur the farm pig by writing words in silk in her web praising the pig. As a result, Wilbur becomes a tourist attraction rather than roast pork.

A LOVE OF ANIMALS

The first inspiration for the story came after E. B. White nursed a sick pig destined for the same fate as Wilbur. The second came after he saw an orb web spider lay an egg sac. Soon after, the female died and he took the egg sac home and watched the babies hatch and eventually disperse. Writing a decade after *Charlotte's Web* was published, White wrote: "Animals are part of my life and I try to report them faithfully and with respect."

← The common barn orb web spider, *Araneus cavaticus* (Araneidae), was the inspiration for the spider in the children's book *Charlotte's Web*.

↙ Charlotte's full name in the book is Charlotte A. Cavatica.

→ The barn spider *Araneus cavaticus*, made famous by *Charlotte's Web*, is a common orb web spider found in North America. Only females build webs. When prey like flying insects become trapped, the female immediately wraps them in a shroud of silk.

SPIDER MYTHS AND MEDICINE

Fortunately, around two thousand years ago, there was a contraceptive available to women that would prevent conception for a year. The advice from Coecailis, a natural historian of the time, was simple: take a large-headed spider and remove two worms from it. Presumably, this advice was confused and the worms were actually the coiled digestive tract in the abdomen. This needed to be wrapped in a piece of deer skin and, before dawn, worn as an amulet. Hopefully, women remembered to renew their prescription a year later, before the prophylactic waned and they risked getting pregnant.

SPIDERS TAKEN INTERNALLY

From the 16th to 19th century in the United Kingdom it seems many people believed a fever could be cured by eating a spider sandwich. Having the magical powers of a spider inside your body was thought to be more effective than simply wearing a spider around your neck.

WEBS AS DRESSINGS

For over two thousand years, spider webs have been used as dressings to cover wounds, just as plasters and bandages are used today. Until recently, it was thought silk had antimicrobial and antifungal properties, which made the wound heal more quickly.

~ Protection from pathogens ~

Recent research, in which the silk from different spiders was tested to see if it had any medicinal benefits, found no evidence that it was able to kill off bacteria or fungi. Some egg sacs are made of very tightly woven silk and this physical barrier may help to protect the eggs inside from pathogens.

DR. MUFFET

Thomas Muffet was an English physician in the latter half of the 16th century. He was very fond of spiders and is often said to be the father of Little Miss Muffet, the famous arachnophobe in the nursery rhyme that first appeared in 1805. This was 200 years after his death and he had no children, so the familial connection is unlikely. Dr. Muffet was cynical of drugs at the time and thought spiders were a better solution for some ailments. He believed gout could be prevented by allowing spiders to build webs inside your house. If you did suffer from gout, a spider wrapped in deerskin and applied to the toe at night would stop the pain.

↓ Seeing spider webs inside a building either means it is abandoned or the owners are happy to live with spiders.

TARANTULAS AT THE MOVIES

Three movies from the 1950s, '60s, and '70s demonstrate three different ways to make tarantulas the villains: make them the size of a house, imply they are unbelievably venomous to people, and make them life-size and hunting in packs.

HUGE TARANTULA

A house-sized tarantula is the biggest star in the 1955 movie *Tarantula*. It's huge because it has been injected with a substance, which in 1950s' science-speak was to "harness the power of the atom." The rampaging spider devours horses and kills a few people before being destroyed by the airforce.

A DEADLY WEAPON

The first James Bond film (starring Sean Connery), *Dr. No* (1962), featured a tarantula being used as a deadly weapon. The two main and extremely well-educated villains, Dr. No and Professor Dent, conspire to use a tarantula to kill their secret agent nemesis. It seems no other weapons are as effective as planting a tarantula in somebody's bed. Of course, Bond kills the spider and eventually the two villains.

THE TARANTULA IN *TARANTULA*

While models of giant tarantula legs were poked into houses to terrify people inside, many scenes in *Tarantula* featured a real spider. To make it look the size of a house, the landscape it is walking through needs to be very small. In the movie, the tarantula was filmed walking across miniature landscapes. Jets of air were blown at the spider to make it walk in the direction the filmmakers wanted. It seems the spider's performance was good enough for it to star in the 1957 movie *The Incredible Shrinking Man*. Here, the spider is life-size and menaces a prey-sized shrinking man.

JAMES BOND AND THE
PINK-TOED TARANTULA

Sean Connery did not like spiders and refused to have a tarantula walking across his body when he was in bed. In a scene where the spider appears to walk across his shoulder and onto a pillow it was walking on glass. You can see it hover, as if in midair, before it steps onto the pillow. A stunt-double was used for the other scenes in which the spider appears to be walking on James Bond. The spider (nicknamed Rosie) was a pink-toed tarantula, *Avicularia avicularia* (Theraphosidae). They are very docile, seldom bite, and their venom is not dangerous to people. Still, it would not have had the same impact if Dr. No had used a stick insect.

↓ A pink-toed tarantula was also cast as a villain in the 2002 movie *Eight Legged Freaks*.

KINGDOM OF THE SPIDERS

While James Bond only had to deal with one tarantula, William Shatner of *Star Trek* fame, and the main star of the 1977 movie *Kingdom of the Spiders*, had to deal with thousands. The spiders are running out of food and have begun hunting in packs so they can catch prey much bigger than a single tarantula could, including people and farm animals. The spiders win when they totally encase the town, where a few survivors are in hiding, in silk webbing.

~ Cast of thousands ~

During the making of *Kingdom of the Spiders*, around 5,000 tarantulas were collected from the desert close to where the movie was made. After the movie was finished the survivors were returned to the desert.

AN ALMOST
VEGETARIAN SPIDER

Jumping spiders are often called "eight-legged cats" because of their excellent eyesight and mammal-like hunting behavior. *Bagheera kiplingi* (family Salticidae) is a jumping spider from Central America, named after the panther in *The Jungle Book* and its author, Rudyard Kipling. Jumping spiders are known to drink nectar from flowers to complement their mostly insect diet.

FOOD-FILLED NODULES

The diet of *B. kiplingi* (Salticidae) is mainly fat, sugar, and protein-rich nodules (known as Beltian bodies) found on the leaf tips of acacia trees. The Beltian bodies are for ants who, in exchange for a home and food, protect the tree from herbivorous browsers. As a neighborhood thief, the spider uses its keen vision to watch out for ants that will attack it for good reasons. They eat the Beltian bodies, their larvae, and nectar-drinking flies and even cannibalize small *B. kiplingi*. The spider is often called a vegetarian spider, although it is more of an opportunistic omnivore.

↓ Ant larvae (species of *Pseudomyrmex*) are common prey for the almost-vegetarian jumping spider *Bagheera kiplingi*.

↓ The leaves and spiky thorns of an acacia branch (species of *Vachellia*), home to the jumping spider *Bagheera kiplingi*.

→ Having grabbed a nutritious Beltian body from the leaf of an acacia tree, the jumping spider *Bagheera kiplingi* sucks up the nodule's nutrients. The spider also drinks nectar and is not totally vegetarian, as it eats the larvae of ants.

SPIDER SUPERPOWERS

On the leaf-littered ground of forests in New Zealand and South America lives a spider with a formidable snap trap, which it uses to stop prey in their tracks.

TRAP-JAW SPIDERS

One trap-jaw spider can snap its jaws shut faster than any other known spider. It has an odd-shaped head with what researchers believe contains the equivalent of tiny rubber bands. When the spider picks up the movement of prey nearby, it raises its closed jaws, so they are horizontal, and opens them wide. This stretches the "rubber bands" and primes the trapping jaw. When hairs on the inside of the jaws are touched, the jaws snap shut in a staggeringly short space of time (see *Snapping at High Speed*, page 125).

PELICAN SPIDERS

If the trap-jaw spider's superpower is speed, a closely related group of similar rice-grain-sized spiders known as pelican or assassin spiders (Archaeidae) have patience as a superpower. Fossils of pelican spiders found in rock and amber suggest they have changed little in around 170 million years. They were known from fossils discovered in the 19th century. Recently, these tiny spiders have been discovered alive and well living on forest floors in South Africa, Madagascar, and Australia.

← Pelican spiders hunt other spiders, and their enormous chelicerae allow them to keep venomous prey at "arm's length" while stabbing.

SNAPPING JAWS

Muscles could never contract fast enough to snap shut the jaws of trap-jaw spiders at such an incredible speed. It is thought the evolution of such fast-snapping jaws was driven by prey that use a type of rubber band to escape from danger. Springtails are tiny insects with a stretched structure folded under their rear abdomen. When this unfolds it pushes down on the ground and launches the insect into the air. In the ongoing evolution of predators catching prey and prey avoiding predators, the snapping of the spider's jaws may stop a pole-vaulting insect from making a spring-loaded escape.

Called pelican spiders because with their long beak they resemble a pelican in profile, they also have a long neck with long chelicerae that rest against it. The spider's mouth is near the fangs at the tips of the jaws. Pelican spiders only eat other spiders. When hunting they follow silken lines on the forest floor left by spiders. If these lead to a potential victim, the pelican spider waits with the patience of an assassin for the right time to strike. Just like the trap-jaw spider, the spider moves its jaws into a horizontal position, before striking and inserting its fangs and venom into the victim. It then removes one of the fangs and waits for the venom to work before bringing its jaws with the impaled insect next to its mouth to begin feeding.

~ Striking at a distance ~

The advantage to the spider in having such long chelicerae is that it keeps prey at a safe distance when caught, especially since their prey is also a predator. Pelican spiders do not hunt other pelican spiders, as they both have weapons with which they could kill each other. Very impressive for a carnivore the size of a grain of rice.

SPIDERS CAN COUNT

J umping spiders are famous for their remarkable eyesight. *Portia* has the best eyesight in this family of over 5,000 species. It's not only their eyesight that is comparable to a small mammal, as they are also far cleverer than an animal with a brain smaller than a pin head should be. Both features may help to explain why *Portia* is so good at catching a variety of its favorite prey: other spiders.

PREDATOR HUNTING PREDATOR

A predator hunting other predators is a dangerous way to find food, but *Portia* has plenty of tricks up its tiny little sleeves. One of these involves spotting prey at a distance, then working out a route to reach it, often with the prey out of sight as *Portia* navigates toward it.

EIGHT-LEGGED CATS

It may seem remarkable that species of *Portia*, a jumping spider with a brain the size of a sesame seed, hunt in a way comparable to the predatory behavior of cats, both big and domestic. However, these arachnid and mammalian predators both use complex behaviors to find, stalk, and catch prey using their excellent eyesight. Spiders were once thought to be simple, hard-wired animals operating on instinct. Decades of research on the predatory strategies and behavioral flexibility of *Portia* have provided insights into the cognitive abilities of a tiny brain and shown this spider is not a hard-wired automaton. In the case of *Portia*, the lights are on and somebody is definitely home.

~ A mental picture ~

Prey may be out of sight, but they are not out of mind. If *Portia* spots a single prey and while the prey is out of sight another prey is added, the spider will stare at the changed number of prey in a different way to when the number of prey is the same. It's as if the spider has a mental representation of the prey it spotted, even if it temporarily loses sight of it.

~ Counting prey ~

Similar to non-verbal human infants, *Portia* can tell the difference between one or two prey. If the number of prey is beyond three, it becomes many. It really is enough to keep a spider biologist awake at night!

← Species of *Portia*, a jumping spider, have been shown to be capble of counting, adding yet another amazingly complex behavior that defies the tiny size of their brains.

ENORMOUS ORB WEB

Cartwheel-shaped orb webs are the most familiar looking spider webs. On the island of Madagascar, Darwin's bark spider, *Caerostris darwini* (Araneidae), builds the largest orb web known, with the widest bridging lines used to suspend the web above rivers, streams, and lakes.

BUILDING THE WEB

The orb web can be almost 32 sq ft (3 sq m) in area with bridging lines 82 ft (25 m) across. The silk used for these bridging lines is the strongest and most elastic produced by any spider. Before building the orb web, the spider sends out numerous lines of bridging silk until it attaches to an anchor on the other side.

CATCHING AQUATIC INSECTS

An aerial trap above wide stretches of water allows the spider to catch prey less likely to be caught by other spiders. In one study, scientists observed around 30 mayflies emerge from the water below and become trapped in a web in a short space of time. That is a lot of excess food and the spider was seen wrapping the prey in silk to keep it fresh for future meals.

~ Freeloaders ~

With so much food on the web, it becomes popular with freeloaders. Flies were seen feeding on the trapped mayflies and the well-known kleptoparasitic spiders—species of *Argyrodes* (Theridiidae)—were also seen stealing food. While the flies can fly in to feast it must be quite a hike for the little kleptoparasite, since it has to walk along the bridge lines before reaching the larder.

← A Darwin's bark spider, *Caerostris darwini*, sits in the middle of her web, waiting to trap prey in the sticky web.

NO VENOM, NO PROBLEM

Unlike most spiders, Uloborid spiders are not venomous, but they can still stop prey in their tracks. Prey are wrapped in hundreds of meters of very fine strands of silk until they are entombed in a silken shroud. A comb on the spider's hind legs combs out the silk, which is about 2,500 times finer than a human hair.

CRUSHING PREY

As the spider wraps, it compresses the prey with ever-tightening silk. This can break its legs and even force its eyes into its head. This reduces the size of the bag of food the spider will need to cover in digestive fluid, before sucking the partially digested liquid through the silken wrapping and into its own digestive system (imagine sucking up soup through a dish towel). Uloborids lost their venom glands because by paralyzing their prey by wrapping instead, these were no longer needed.

↓ The cribellate orb weaver *Uloborus walckenaerius* (Uloboridae) is known as the feather-legged spider and has long front legs.

↓ By hiding in plain sight on a silken stabilimentum in its web, *U. walckenaerius* can be less obvious to potential predators.

→ Sitting in her web is a female little humped spider, *Philoponella congregabilis* (Uloboridae). While the wrapped parcel below the spider looks like a large disguised egg sac, it is, in fact, prey that has been trapped in a thick wrapping of very fine silk.

SOCIAL SPIDERS

M ost spiders lead solitary lives and only engage with their own species when mating or cannibalizing. Around 20 species (out of around 50,000 species of spider) have embraced communal living, creating the spider equivalent of fish schools and mammal herds.

LIVING TOGETHER

The collective noun for a group of spiders is a cluster, and social spiders are usually found living in areas with a large number of prey, so there is enough food for the colony. The most studied social spider is *Anelosimus eximius* (Theridiidae) from South America. While the spiders are tiny (around 5/16 in/8 mm), their communal webs can be 26 ft (8 m) wide and 5 ft (1.5 m) high, with thousands of spiders living in a single colony. The spiders hunt in packs, capturing prey much larger than would be possible for an individual spider.

AFRICAN SOCIAL SPIDERS

Stegodyphus dumicola (Eresidae) from arid regions of Central and Southern Africa is commonly known as the African social spider. The spiders in the colony cooperate to catch prey in their web and are able to catch insects much larger than an individual spider could. The African social spider lives in colonies of only a few hundred

individuals rather than the thousands found in the colonies of their South American counterpart. Within these smaller colonies, about 80 percent of the spiders are female and less than half of the females breed.

← When the African social spider, *Stegodyphus dumicola*, catches large prey, it partially digests it, before sharing the meal with the colony.

HUNTING IN A PACK

Recent research using high-speed cameras to film *Anelosimus eximius* (Theridiidae) spiders hunting has shown they move and pause collectively for a fraction of a second. When paused the spiders can "listen" for where the silk-borne vibrations from prey are coming from, without the background noise from the vibrations of the group walking across the web. The most common insects caught by the spiders are ants, beetles, and cockroaches. Occasionally, large crickets and moths, which can weigh up to 700 times the weight of an individual spider, become trapped in the web and can be captured by the pack. Being a member of a cluster of spiders is much more than just safety in numbers.

However, virgin females still cooperate in the care of offspring, despite not being the spiderlings' real mothers. They care for and guard the egg sac and regurgitate food for the spiderlings once they have hatched, in a way similar to birds regurgitating food for their chicks.

~ A mother's gift ~

Once the baby spiders have molted a few times, the mothers and virgin females provide a final additional meal for the offspring: themselves. In what seems like an extremely gruesome act the females deliberately allow the offspring to feed on them. In fact, they prepare their bodies for the developing spiders to feed on by liquifying their abdomens. They keep essential organs intact, so they are still alive when they are eaten. Since the spiders in the colony are so closely related, non-breeding females still help pass their genes onto the next generation because of their relatedness to the offspring. Also, it has been shown that the contribution by the virgin females produces bigger and healthier offspring.

GLOSSARY

abdomen
The rear part of a spider's body.

anterior aorta
A large artery that takes blood from the heart to the cephalothorax.

araneophagy
Foraging tactic whereby spiders eat other spiders.

autotomy
A mechanism that allows spiders to self-amputate one or more legs.

beltian body
A nutritious nodule found on the leaves of acacia trees and eaten by a species of jumping spider.

book lungs
Breathing organs found on the underside of the abdomen.

cephalothorax
(fused head and thorax) The front part of a spider's body to which the legs and mouthparts are attached.

chelicera
(pl. chelicerae) Jaw located at the front of the cephalothorax. The jaws contain the basal segments and the fangs.

coxa
Attached to the cephalothorax, this is the first joint of a spider's seven-segmented leg.

egg sac
A silken sac used to protect and hold spider eggs until they hatch.

epigynum
A genital opening on the underside of the female's abdomen, into which the male inserts his palp and transfers sperm.

exoskeleton
The external skeleton of a spider, consisting of the rigid cephalothroax, legs, and mouthparts as well as the more flexible abdomen.

hemocyanin
The oxygen-carrying molecule found in spider blood.

hemolymph
Spider blood that carries nutrients and oxygen around the body and also drives a hydraulic mechanism essential for walking.

lyriform organ
Sense organ on the spider's legs that picks up vibrations on the ground.

malpighian tubule
Organ that filters waste and conserves water, similar to our kidneys.

matriphagy
When baby spiders feed on their mother's body.

metatarsus
The second-to-last joint on a spider's seven-segmented leg.

pedicel
A narrow, circular waist that connects the cephalothorax to the abdomen.

pedipalps
A pair of short, leg-like structures located either side of the chelicerae. They are used to manipulate prey and, in males, to carry sperm.

posterior aorta
A large artery that takes blood from the heart to the abdomen.

spermatheca
Special package inside the female used to store sperm from the male.

spiderling
A young spider that has recently hatched from an egg.

spigots
Tiny, tube-like pores on the spinnerets, which contain liquid silk that is pulled out as a silken thread.

spinnerets
(silk-spinning organ)
Located at the rear of the abdomen, these finger-like projections are used by the spider to control the use of silk.

spiracle
An opening that connects to breathing tubes called tracheae.

stabilimentum
(pl. stabilimenta)
A silken decoration on a web, like a spiral of silk, that may stop birds from flying into the web.

tapetum lucidum
A structure that reflects light that has passed through the retina back onto the retina to give better vision in low light.

tarsus
The last segment of a spider's leg, which has tarsal claws and the sensory tarsal organ.

tracheal system
A highly branched series of breathing tubes that open into one or two spiracles.

trichobothria
Sensory hairs found on spider legs that can pick up airborne vibrations.

trochanter
The second joint of a spider's seven-segmented leg.

FURTHER READING

Foelix, R. 2011. *Biology of Spiders*, 3rd edition. NY: Oxford University Press.

Herberstein, M. (editor). 2011. *Spider Behaviour: Flexibility and Versatility*. UK: Cambridge University Press.

Nelson, X. 2024. *The Lives of Spiders: A Natural History of the World's Spiders*. NJ: Princeton University Press.

Platnick, N. (editor). 2020. *Spiders of the World: A Natural History*. NJ: Princeton University Press.

Pollard, S. and P. Sirvid. 2021. *Why is that Spider Dancing? The Amazing Arachnids of Aotearoa*. Te Papa Press, Wellington, New Zealand.

INDEX

ACKNOWLEDGMENTS

Thank you to UniPress Books and its publisher, Nigel Browning, for asking me to write *The Little Book of Spiders*. I could not have wished for a better project and thank you to Ruth Patrick and Lindsey Johns for their fantastic support and help with the book and its illustrations. Also thank you to Tugce Okay for her elegant and timeless paintings that make this such a special book. I also want to thank my fellow kiwi spider biologists in New Zealand (yes, there are a few of us), Robert Jackson, Fiona Cross, Phil Sirvid, Cor Vink, Ximena Nelson, and Vicki Smith for sharing their knowledge and stories. Finally, a big thank you to my wonderful wife, Cynthia Cripps, for her support during this project and for being a fantastic sounding board for the spider stories that shaped this book.

ABOUT THE AUTHOR

Simon D. Pollard grew up in Christchurch, New Zealand. From the age of seven Simon wanted to study animals at university. At 20 he became fascinated with spiders while at the University of Canterbury, where he graduated with a PhD. Simon has been involved in spider research projects in New Zealand, Asia, Africa, and North America. He makes his research and the world of natural history accessible to a wide audience through talks, and the writing of popular articles and books for adults and young readers. Since 2009 he has been Adjunct Professor of Science Communication at the University of Canterbury.